Multi-Sided Platforms (MSPs) and Sharing Strategies in the Digital Economy:

Emerging Research and Opportunities

Sergey Yablonsky
St. Petersburg University, Russia

A volume in the Advances in E-Business Research (AEBR) Book Series

IGI Global
DISSEMINATOR OF KNOWLEDGE

Published in the United States of America by
IGI Global
Business Science Reference (an imprint of IGI Global)
701 E. Chocolate Avenue
Hershey PA, USA 17033
Tel: 717-533-8845
Fax: 717-533-8661
E-mail: cust@igi-global.com
Web site: http://www.igi-global.com

Library of Congress Cataloging-in-Publication Data

Names: Yablonsky, Sergey, 1953- author.
Title: Multi-sided platforms (MSPs) and sharing strategies in the digital economy : emerging research and opportunities / by Sergey Yablonsky.
Description: Hershey : Business Science Reference, [2018]
Identifiers: LCCN 2017045860| ISBN 9781522554578 (hardcover) | ISBN 9781522554585 (ebook)
Subjects: LCSH: Information technology--Economic aspects. | Digital electronics.
Classification: LCC HC79.I55 Y33 2018 | DDC 658.8/72--dc23 LC record available at https://lccn.loc.gov/2017045860

This book is published in the IGI Global book series Advances in E-Business Research (AEBR) (ISSN: 1935-2700; eISSN: 1935-2719)

British Cataloguing in Publication Data
A Cataloguing in Publication record for this book is available from the British Library.

Advances in E-Business Research (AEBR) Book Series

ISSN:1935-2700
EISSN:1935-2719

Editor-in-Chief: In Lee, Western Illinois University, USA

MISSION

Technology has played a vital role in the emergence of e-business and its applications incorporate strategies. These processes have aided in the use of electronic transactions via telecommunications networks for collaborating with business partners, buying and selling of goods and services, and customer service. Research in this field continues to develop into a wide range of topics, including marketing, psychology, information systems, accounting, economics, and computer science.

The **Advances in E-Business Research (AEBR) Book Series** provides multidisciplinary references for researchers and practitioners in this area. Instructors, researchers, and professionals interested in the most up-to-date research on the concepts, issues, applications, and trends in the e-business field will find this collection, or individual books, extremely useful. This collection contains the highest quality academic books that advance understanding of e-business and addresses the challenges faced by researchers and practitioners.

COVERAGE

- Mobile Business Models
- Valuing e-business assets
- E-business standardizations
- Trust, Security, and Privacy
- E-business models and architectures
- Electronic communications
- Outsourcing and e-business technologies
- E-procurement methods
- Web Service-Based Business Systems
- Web advertising

IGI Global is currently accepting manuscripts for publication within this series. To submit a proposal for a volume in this series, please contact our Acquisition Editors at Acquisitions@igi-global.com or visit: http://www.igi-global.com/publish/.

The Advances in E-Business Research (AEBR) Book Series (ISSN 1935-2700) is published by IGI Global, 701 E. Chocolate Avenue, Hershey, PA 17033-1240, USA, www.igi-global.com. This series is composed of titles available for purchase individually; each title is edited to be contextually exclusive from any other title within the series. For pricing and ordering information please visit http://www.igi-global.com/book-series/advances-business-research/37144. Postmaster: Send all address changes to above address.

Titles in this Series

For a list of additional titles in this series, please visit:
https://www.igi-global.com/book-series/advances-business-research/37144

Mobile Platforms, Design, and Apps for Social Commerce
Jean-Éric Pelet (ESCE International Business School, Paris, France)
Business Science Reference • ©2017 • 411pp • H/C (ISBN: 9781522524694) • US $215.00

The Internet of Things in the Modern Business Environment
In Lee (Western Illinois University, USA)
Business Science Reference • ©2017 • 340pp • H/C (ISBN: 9781522521044) • US $205.00

Key Challenges and Opportunities in Web Entrepreneurship
Alexandru Capatina (Dunarea de Jos University of Galati, Romania) and Elisa Rancati (University of Milan – Bicocca, Italy & Beta Consulting, Italy)
Business Science Reference • ©2017 • 291pp • H/C (ISBN: 9781522524663) • US $205.00

Apps Management and E-Commerce Transactions in Real-Time
Sajad Rezaei (Taylor's University, Malaysia)
Business Science Reference • ©2017 • 379pp • H/C (ISBN: 9781522524496) • US $210.00

Driving Innovation and Business Success in the Digital Economy
Ionica Oncioiu (Titu Maiorescu University, Romania)
Business Science Reference • ©2017 • 321pp • H/C (ISBN: 9781522517795) • US $200.00

Social Media Listening and Monitoring for Business Applications
N. Raghavendra Rao (FINAIT Consultancy Services, India)
Business Science Reference • ©2017 • 470pp • H/C (ISBN: 9781522508465) • US $205.00

Analyzing the Strategic Role of Social Networking in Firm Growth and Productivity
Vladlena Benson (University of West London, UK) Ronald Tuninga (Kingston Business School, UK) and George Saridakis (Kingston University, UK)
Business Science Reference • ©2017 • 525pp • H/C (ISBN: 9781522505594) • US $220.00

For an entire list of titles in this series, please visit:
https://www.igi-global.com/book-series/advances-business-research/37144

IGI Global
DISSEMINATOR OF KNOWLEDGE

701 East Chocolate Avenue, Hershey, PA 17033, USA
Tel: 717-533-8845 x100 • Fax: 717-533-8661
E-Mail: cust@igi-global.com • www.igi-global.com

Table of Contents

Preface vi

Chapter 1
Digital Platforms: Definitions, Strategy, and Business Models 1

Chapter 2
Digital Platforms: From Technology to Leadership Platforms 44

Chapter 3
Innovation Platforms: Data and Analytics Platforms 72

Chapter 4
Innovation Platforms: IOT Platforms 96

Chapter 5
Transaction Platforms: Fintech Platforms 113

Chapter 6
Integrated and Transactional Platforms: Search Platforms 134

Chapter 7
Integrated and Transactional Platforms: Crowdsourcing and Crowdfunding Platforms 146

Conclusion 167

Related Readings 170

Index 191

Preface

Digital technologies, including the Internet and other related technologies, are changing the way the companies do business and are the principal driver of the surge of interest in business models. Over recent decades there has been a rapid global transition from industrial economics to digital economy (Judah, 2016). As defined by Gartner, digital economy examines the creation, consumption and control of value associated with digital products, services and assets in organizations to drive for new business models and create new markets (Judah, 2016). Digital economy have a relatively high frequency of acquiring neologisms (Kornberger, Pflueger, Mouritsen, 2017), including platform capitalism (Srnicek, 2016), multi-sided markets (Evans & Schmalensee, 2016), sharing economy (Sundararajan, 2016), collaborative consumption (Botsman & Rogers, 2010), gig economy (Mulcahy, 2016), mesh (Gansky, 2010) or commons-based peer production (Benkler, 2011).

Digital economy will force enterprises to rethink how they deal with economic agents that can be a person, a business, a thing, a smart machine or even an algorithm. Frequently an industry's transformation is associated with the adoption of a new technology. But even though new technologies indeed mostly act as major factors, they have never transformed an industry on their own. Kavadias, Ladas and Loch (2016) showed that what does achieve such a transformation is a business model that can link a new technology to an emerging market need. Evans and Gawer (2016) proved that today the platform business model can be found in a growing number of industries including social networking (Facebook, LinkedIn); Internet auctions and retail (Amazon, eBay, Angie's List); on-line financial and human resource functions (Workday, Elance-oDesk, Freelancer, WorkFusion), urban transportation (Uber, Lyft, Sidecar), mobile payment (Mahala, Square) and clean energy (Sungevity, SolarCity, EnerNOC).

In a digital business ecosystem main economic agents (humans, business and smart things/smart machines) can play any combination of roles, such as customer, partner, supplier, employee or competitor. The systems that support digital businesses require transformation to maximize the economic value of network connections. It forces a shift to platforms and ecosystems collaborative by nature, by applying business models with multiple actors playing multiple roles. The rise of platforms as new organizational form includes accommodation providers such as Airbnb, ride-sharing companies such as Uber, service and product marketplaces such as Taskrabbit or eBay, and even relationship services such as eHarmony. Truly, there is an Uber for Everything Now as the Wall Street Journal commented in 2015 (Fowler, 2015). The platform ecosystem needs to accommodate new technologies easily as dynamic business capabilities arise (Teece, 2017). Digital business obligates companies to evolve their business models as new opportunities arise (Scheibenreif, 2016). Therefore, enterprises will need to design for flexibility with business models and the metrics that measure their performance. (Scheibenreif, 2016).

Faster digitalization reveals the growing gap between current economic terminology understanding and new digital business models, depriving researchers of a holistic view of business value. For example, Teece (2017) mentions that today the concept of "industry" as a group of firms performing similar activities and competing or cooperating with each other is less and less aligned with the way firms think about themselves. In the digital economy, continues Teece (2017), firms see their role less in industries and more in business ecosystems. Business ecosystems consist of organizations and customers working together to create and sustain markets, products and services. Scheibenreif (2016) argues that digital economy strategy should permit the organization to estimate technology investment decisions and to foster flexible methods for adapting to a more dynamic business environment.

The coevolution of the system is typically reliant on the technological leadership of one or two firms that provide a platform around which other system members who offer inputs and complementary goods, align their investments and strategies. Some ecosystems may involve multiple platforms at different levels of the value chain. A platform provides common standards, interfaces and tools to leverage core technologies in order to increase the productivity and profitability of a company, a set of companies or users. The term was initially used in relation to internal product platforms. The definition was then expanded to apply to the settings where one company provides a hub that other companies utilize in their own businesses (Gawer and Cusumano, 2002).

Faster digitalization reveals the growing gap between current economic value indicators and new digital platform business models, depriving researchers of a holistic view of business value.

This study aims to systematize the main areas of research emerging from the intersection of digital economics, technology and open innovations to establish a future research agenda and create a platform framework for organizations to understand and account for how much of their organization's business value is digital value, complementing traditional value derived from products and services.

Innovative network infrastructure will foster new platform strategies and platform business models suited to a digital business and commerce activity where life becomes digital. An intelligent infrastructure that will merge the physical and virtual worlds is just starting to be built. Gartner argues that digital business is the creation of new business designs by blurring the digital and physical worlds. Managers will have to help their enterprise contribute to the development of this digital business and commerce. Digital commerce is the buying and selling of goods and services using digital channels and commerce infrastructure (Yablonsky, 2016). It contains marketing activities that support these transactions, including people, processes and technologies to execute the offering of development content, analytics, promotion, pricing, customer acquisition and retention, and customer experience at all touchpoints throughout the customer buying journey (Gartner, 2017).

Multi-sided platforms (henceforth, MSP) comprise a large and rapidly growing share of the world digital economy. MSPs are increasingly prevalent: 60 of the world's 100 largest global companies earn most of their revenue from MSPs. Today every organization needs a digital platform strategy (Burton et.al, 2016). It must collaborate with their business counterparts to integrate digital platform development into their technology and business strategic planning.

This research discusses platform strategy and business models in a general way and MSP business models in particular. It provides a broad overview of platforms, specifies platform taxonomy, proposes and examines main business models and cases. Business models have received significant attention from both practitioners and academics (Gassmann et al., 2016). A good representation of MSP business models can help to illuminate important dimensions of a problem – in our case, to understand and classify digital platform intermediaries across different markets. We illustrate, analyze and classify platforms by means of ontologies, in particular lightweight ontologies – taxonomies.

Our research was motivated by the observation that various descriptions of digital platforms implement the concept of a multi-sided platform business model on similar core concepts with distinct features. Having a common language in turn makes it easier to visualize digital platforms using a common set of representation techniques. Thus, we illustrate how the platform innovative taxonomy can be complemented by the real instances in the global and emerging markets.

Figure 1. Book structure

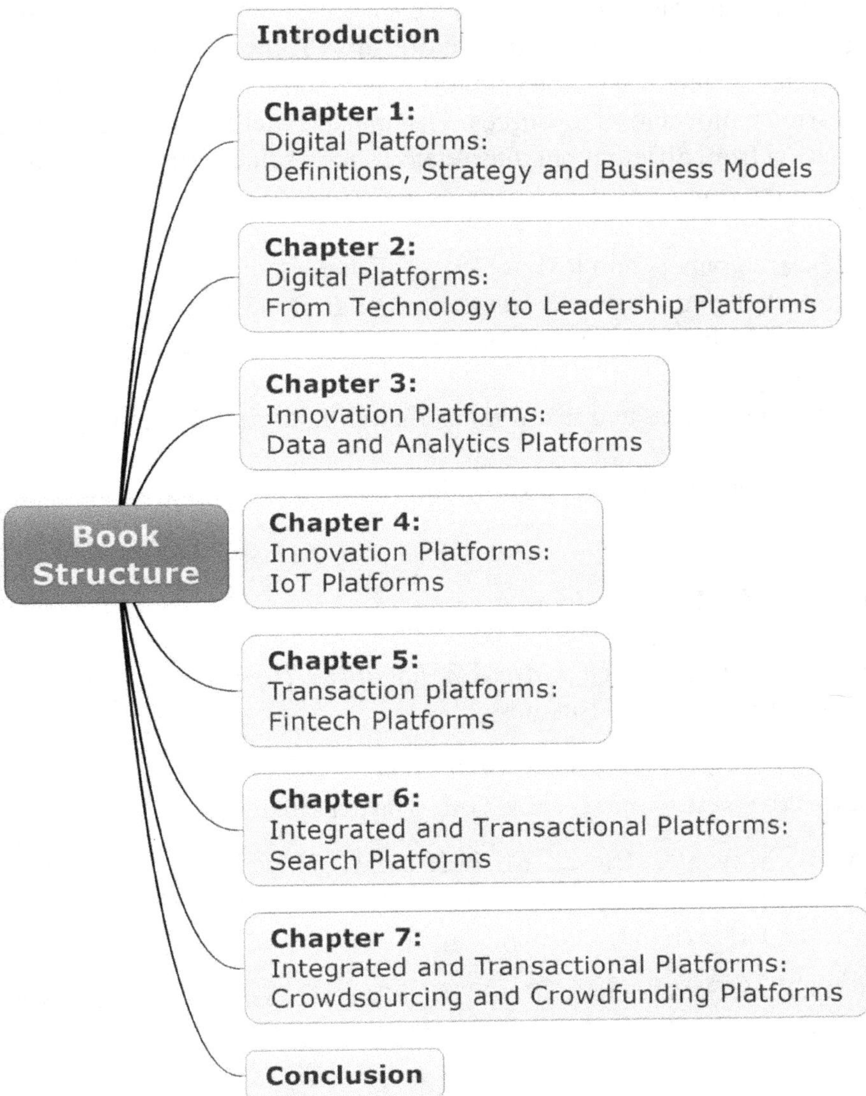

This book conducts a systematic mapping study on platform intermediaries in digital business and commerce research. The presented qualitative conceptual research aims to collect and analyze quality data regarding the current status and prospective evolution of platform intermediaries offered by leading digital business companies that are running businesses in different industry sectors. In our research, we argue that, rather than searching for a single acceptable definition, a better approach would be to develop a classification system or taxonomy (Poli et al. 2010).

Taxonomies and other types of controlled vocabularies are the preferred means to achieve such a common understanding by specifying the terms of the domain, disambiguating them from each other, controlling synonyms and structuring the domain via term relationships (Yablonsky 2014, 2016). For a conceptual grounding of the categories in taxonomies, we use definitions from various information resources. The analysis methodology incorporates information from different but interrelated sources that form a representative sample of the domain:

- Research papers and text books, published from 2001 up to 2017;
- Reported and observed trends and digital business and commerce activity. This study incorporates paper/Internet published data from 2001 up to 2017;
- Open ontologies and taxonomies including digital domain.

The brief book structure is shown on Figure 1.

REFERENCES

Benkler, Y. (2011). *The penguin and the leviathan: How cooperation triumphs over self interest*. Crown Business.

Botsman, R., & Rogers, R. (2010). *What's mine is yours. The rise of collaborative consumption*. New York: Harper Business.

Burton, B., Moyer, K., Blosch, M., LeHong, H., Howard, C., & Scheibenreif, D. (2016). Every Organization Needs a Digital Platform Strategy. Gartner Report, ID: G00316151.

Evans, D. S., & Schmalensee, R. (2016). *Matchmakers: The New Economics of Multisided Platforms*. Harvard Business Review Press.

Evans, P. C., & Gawer, A. (2016). *The Rise of the Platform Enterprise A Global Survey.* New York: Center for Global Enterprise. Retrieved December 18, 2018, from: http://www.thecge.net/wp-content/uploads/2016/01/PDF-WEBPlatform-Survey_01_12.pdf

Fowler, G. (2015). *There's an Uber for Everything Now. Apps do your chores: shopping, parking, cooking, cleaning, packing, shipping and more.* Retrieved December 18, 2018, from: https://www.wsj.com/articles/theres-an-uber-for-everything-now-1430845789

Gansky, L. (2010). *The mesh: Why the future of business is sharing.* Penguin.

Gartner. (2017). *Gartner IT Glossary.* Retrieved December 18, 2018, from: http://www.gartner.com/it-glossary/

Gassmann, O., Frankenberger, K., & Sauer, R. (2016). *Exploring the Field of Business Model Innovation. New Theoretical Perspectives.* Palgrave Macmillan.

Gawer, A., & Cusumano, M. (2002). *Platform Leadership: How Intel, Microsoft and Cisco Drive Industry Innovation.* Harvard Business School Press.

Judah, S. (2016). Disruption and the Digital Value of Business — the Rise of Digital Economics. Gartner Report, ID: G00311450.

Kavadias, S., Ladas, K., & Loch, C. (2016, October). The Transformative Business Model. How to tell if you have one. *Harvard Business Review,* (10), 91–98.

Kornberger, M., Pflueger, D., & Mouritsen, J. (2017). *Evaluative infrastructures: Accounting for platform organization, Accounting, Organizations and Society.* Academic Press. (in press)

Mulcahy, D. (2016). *The Gig Economy: The Complete Guide to Getting Better Work, Taking More Time Off, and Financing the Life You Want.* New York: Amacom.

Poli, R., Healy, M., & Kameas, A. (2010). *Theory and Applications of Ontology: Computer Applications Set: Theory and Applications of Ontology.* Springer.

Scheibenreif, D. (2016). *Unlock Digital Business Value Through the Economics of Connections.* Gartner Report ID: G00293686.

Srnicek, N. (2016). *Platform capitalism.* Polity.

Sundararajan, A. (2016). *The sharing economy*. Cambridge, MA: MIT Press.

Tauscher, K., & Laudien, S. (2017). Understanding platform business models: A mixed methods study of marketplaces. *European Management Journal*. (in press)

Teece, D. (2017). Business models and dynamic capabilities. *Long Range Planning*. (in press)

Yablonsky, S. (2014). ICT-based Financial Innovations. In A.-L. Mention & M. Torkkeli (Eds.), *Innovation in financial services: a dual ambiguity* (pp. 184–224). Cambridge Scholars Publishing.

Yablonsky, S. (2016). Intermediaries in E-Commerce. In *Encyclopedia of E-Commerce Development, Implementation, and Management* (pp. 48–73). IGI Global. doi:10.4018/978-1-4666-9787-4.ch005

Chapter 1
Digital Platforms:
Definitions, Strategy, and Business Models

ABSTRACT

This chapter reviews the following key aspects of platform research: platform strategy, dynamic capabilities, and business models. The main platform typologies and basic definitions are described first. It provides a brief summary of the literature relating to arguing platform strategy, platform life cycle, platform building blocks, and business models. A platform strategy categorization taxonomy and platform practical strategies of preventing platform disintermediation are developed. The main types of platform business models are introduced. The multi-sided platform business model pattern (MSP BMP) is designed. MSP BMP is used as a basic conceptual framework and knowledge management tool for describing, analyzing, and interpreting non-price instruments used by digital platforms, especially platform intermediaries.

DEFINING PLATFORMS

Platforms have lasted for years: offline malls link consumers and merchants; paper newspapers connect subscribers and advertisers, etc. As stated in a global survey, now platforms have become an important economic force with a total market value of $4.3 trillion and an employment base of at least 1.3 million direct employees and millions of others indirectly employed (Evans and Gawer, 2016).

DOI: 10.4018/978-1-5225-5457-8.ch001

Platforms are increasingly prevalent: sixteen of the twenty-five most valuable brands for 2014 function as platforms (Taube, 2014). The list includes Visa and MasterCard, the credit card companies which function as two-sided platforms and facilitate the interaction between cardholders and merchants; Facebook, which started as a digital one-sided platform (enabling the interaction between one distinctive group of university users) and developed into being a multi-sided platform (connecting users, advertisers and third-party game content providers); Amazon, which adopts a multiple business model functioning as a platform when it offers products that are sold by third-party sellers and as a reseller when it sells items under its own name (e.g. Amazon Kindle) (Hagiu, 2006).

Among the largest IPOs of the last few years, many are multi-sided platforms: Alibaba, the biggest IPO ever, with $25 billion; Visa, $19.7 billion; Facebook, $16 billion; Twitter, $2.1 billion; Google, $1.67 billion; LinkedIn, $1.2 billion; Groupon, $700 million. Airbnb or Über threaten to disrupt well-established industries like taxi or hospitality services using platform as a business model. It is stated that platforms facilitate generation of a potentially very large number of complementary innovations by tapping into the innovative capabilities of many external actors, and function as a technological foundation at the heart of innovative business ecosystems (Tiwana, 2014). Platforms are functioning in Europe, North America, Asia, Africa and Latin America. Evans and Gawer (2016) have shown that platform companies are now a visible global phenomenon: they are discovered not only in advanced industrial markets, but throughout the entire world thanks to the growing availability of mobile digital technologies and Internet.

Study examines platform innovations, platform strategies, platform business models, and the cases that illustrate how these business models could be successfully implemented.

We should distinguish a technology platform from a business platform, which includes technology platform as a main block.

There are different definitions of the term technology platform (Gawer and Cusumano, 2002; 2008; Gawer, 2009; Hagiu, 2014; Evans and Gawer, 2016).

Definition 1. (Gawer, 2009)

A *technology platform* is defined as building blocks that act as a foundation upon which an array of firms (a business ecosystem) develop complementary products, technologies or services, proposing such requirements for a platform:

- It should perform a critical function of the overall system or should solve a crucial technological issue of an industry,
- It should be easy to connect to, ''build upon'' and provide space for new and unplanned usage.

There are two main groups of technological platforms: *product/service platforms* and *network platforms*.

Platforms serve as the foundation for complementary products and services (Cusumano and Gawer, 2002). Robertson and Ulrich (1998) initially used the term "platform" in relation to internal product platforms such as automobile chassis or unit-bodies, which permitted multiple products to be generated around the same basic inputs. The definition was then expanded to apply to settings where one company provides an offline/online hub that other companies utilize in their own businesses (Gawer and Cusumano, 2002) and was widely used for Internet portal providers. More recently, it has been applied to many-sided market infrastructures (Iansiti and Levien, 2004).

Defenition 2. (Parker and Van Alstyne 2015).

A *network technological platform or network platform* is usually based on digital network(s) building blocks. Network platforms differ from product platforms because of stronger network effects, switching costs and multi-homing costs.

Definition 3. (Eisenmann et al., 2011).

A *business platform* is a nexus of rules and infrastructure that facilitates interactions among network users.

Definition 4. (Parker and Van Alstyne, 2015).

A *business platform* is a published standard, together with a governance model, that facilitates third-party participation.

Platforms have unique characteristics, with one of the central features being the presence of network effects (Shapiro, Varian, 1998; Rochet and Tirole, 2003; Parker and Alstyne, 2005; Eisenmann, Parker, Alstyne, 2006; Hagiu and Rothman, 2016). The value of platform affiliation for a user depends on the number of other users.

Platforms match buyers with suppliers who transact directly with each other using system resources (Hagiu, 2006; Hagiu and Wright, 2015a,b; Parker and Van Alstyne, 2015) and are generally subject to network effects (Eisenmann, 2007; 2011; Boudreau, 2010).

Most of today's platforms are principally digital. They capture, transmit and monetize data, including personal data, over the Internet and other networks (Evans and Gawer, 2016). They may not be purely digital, but most successful digital platforms currently take advantage of the digital innovations and power of pervasive Internet connectivity in the hand of billions of users. Platform technologies are essential for coordinating suppliers and buyers to deliver consumer value. Facebook is a type of platform that coordinates suppliers and consumers of social information. Without a social media platform such as Facebook, it would be much more difficult and costly for people to exchange social information.

Industries can contain competing platforms. Globally Facebook competes with Google Plus and Twitter. In some regions local platforms could compete with well-known global platforms. For example, in Russia a social media platform VK is successfully competing with Facebook, a search engine platform Yandex with Google. In China a local search engine Baidu dominates against Google. In the video gaming industry, Sony's PlayStation competes with Microsoft's Xbox and Nintendo's Wii. However, some industries have seen the emergence of industry platforms that shape their technological evolution. The firms that own such platforms are *platform leaders*.

Not every new technology has a platform potential, but every firm have to understand the importance of platforms. Companies with a product that has platform potential need a strategy to win a platform war (which is analogous to a standards war) without giving up too much value. Firms producing products that are part of an ecosystem dominated by an industry platform need a strategy to remain compatible with that platform while avoiding commoditization or competition from the platform owner.

Definition 5 (Teece, 2012).

A *business platform ecosystem* contains a number of firms and other institutions that work together to create and sustain new markets and new products. An ecosystem may be anchored by a platform. The co-evolution of the ecosystem is typically reliant on the technological leadership of one or two firms that provide a platform around, which other system members that provide inputs and complementary goods align their investments and strategies with.

Figure 1. Platform typology (adopted from Evans and Gawer, 2016)

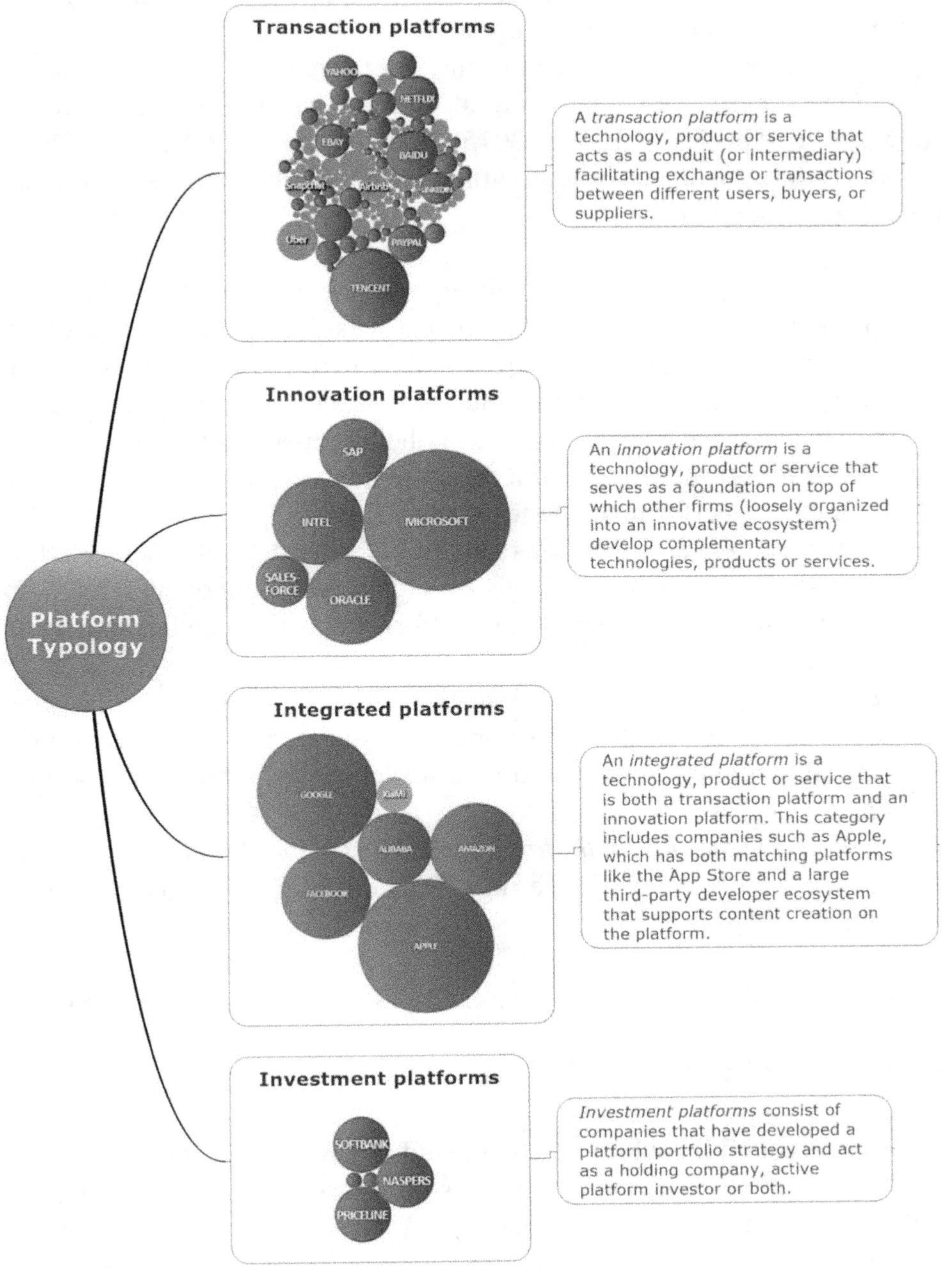

A platform provides common standards, interfaces and tools to power core technologies in order to increase the productivity and profitability of a company, a set of companies, or users (Teece, 2017a,b).

Platforms can take a variety of forms. There are many ways to categorize them. One is based on the structure of platform ecosystem and control. As shown by Parker, Van Alstyne (2015), Alstyne, Parker, Choudary (2016), platforms all have an ecosystem with the same basic structure, comprising four types of actors/player/roles:

1. *Owners* of platforms or platform *sponsors* control their intellectual property and governance. The owner sets the direction, controls the underlying platform technology and provides the overall organizing structure for the platform via rules, governance, and ecosystem support. It can assist the ecosystem work by helping participants see how they are better off by being part of the system rather than outside of it. This role can be performed by one or many firms.
2. *Platform providers* operate as a platforms' interface with users. Platform provider is the point of contact for common components, rules and architecture for the users of the platform. This role can be done by one or many firms.
3. *Users (supply side)* or *the producers* create their offerings. These are content and application developers. Producers provide specific items that attract users to the platform – music, games, information, services etc.
4. *Users (demand side)* or *the target consumers* of the platform solutions and services use producers' offerings. They can be individuals, businesses, organizations.

Another platform typology (Evans and Gawer, 2016) is shown on Fgure 1.

Platform Building Blocks

Platform is usually a subset of components (hardware, software, services) and rules (technical standards, protocols for information exchange, policies, and contracts that govern transactions) employed by users in most of their transactions. All this forms main multi-sided platform (henceforth, MSP) building blocks (Figure 2).

Figure 2. Platform building blocks

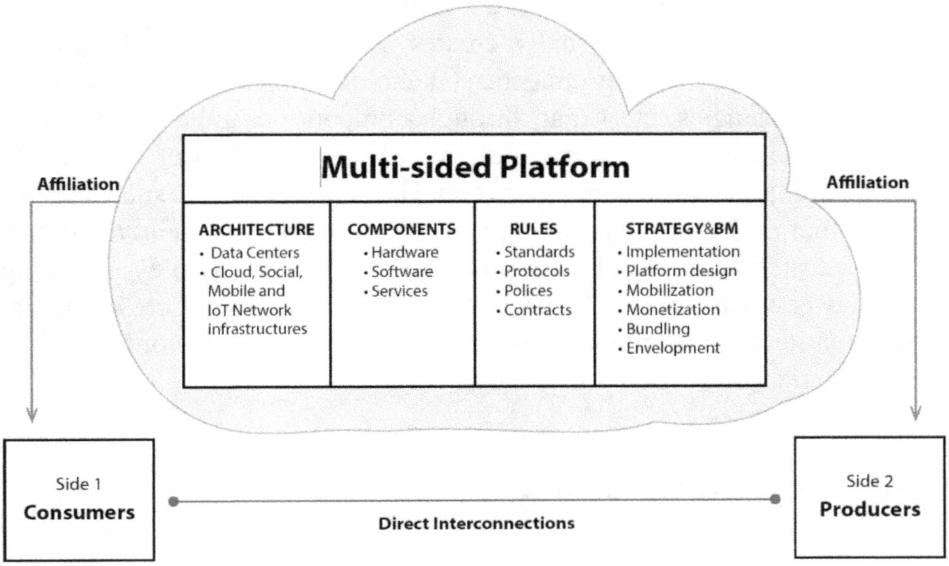

For example, eBay's platform has the following main components and rules (Eisenmann, 2007):

- Components
    - Bid tracking software
    - Shipping services (by 3rd parties)
    - Links to PayPal (an eBay-owned platform)
    - Browser and Internet access (by 3rd parties)
    - Website design
    - Other
- Rules
    - Registration requirements
    - Dispute resolution processes
    - Feedback system
    - Policies for bidding
    - Other

Platforms enable new products/services by reusing platform components. They have lower fixed costs and enable service providers to market in a shorter period of time. Market intermediaries usually reduce search and transactions costs for interactions among two or more distinct groups of customers (e.g. suppliers and consumers for 7-Eleven; sellers and buyers for eBay).

A successful platform usually creates a network of the relationships among economic entities (producers, stakeholders, distributors, consumers, government agencies, etc.) that, through competition and/or cooperation, facilitate the creation and distribution of a platform product or service. The platform digital ecosystem metaphor could be used to describe such platform network that is characterized by open, flexible, demand-driven, interactive networked architecture and collaborative environments (Boley & Chang, 2007; Gawer & Cusumano, 2014). For some social MSPs and crowdfunding MSPs creation of the digital ecosystem is a critical success factor for positive MSP dynamics.

PLATFORM VALUE CHAIN

How platforms differ from the conventional "pipeline" businesses that have dominated industry for decades? Pipeline businesses create value by controlling a linear series of activities — the classic value-chain model. Inputs at one end of the chain (say, materials from suppliers) undergo a series of steps that transform them into an output that is worth more: the finished product (Alstyne, Parker, Choudary, 2016), and value is added along the way.

Among the existing definitions, each one points out that reconfiguration of platform assets allows external parties to interact with each other and add value. Examples include operating systems, game consoles, payment systems, ride-sharing platforms, smart grids, healthcare networks and social networks.

Digital economy changes the value-chain understanding. In the economy of connections, a company's success depends on it's role in one or more networks (Blaschke, Haki1, Riss, Aier, 2017). Digital technologies inforce a paradigmatic shift in economic exchange from a goods-dominant (G-D) to a service-dominant (S-D) logic (Vargo, Lusch, 2016; Blaschke, Haki1, Riss, Aier, 2017). Service-dominant logic needs and facilitates coordination and cooperation involved in the co-creation of value through markets and, more broadly, in the society.

Parker and Van Alstyne (2015) stated that building on Thomson's (1967) typology of long-linked, mediating and intensive technologies, Stabell and Fjeldstad (1998) identified network platforms as one of the three elemental configurations through which firms generate value. In traditional industries bilateral exchanges follow a linear path as companies' purchase inputs, transform them to add value, assemble components and subsystems into complete products and then sell the output.

A platform provides the infrastructure and rules for a marketplace that brings together producers and consumers. The players in the ecosystem fill four main roles but may shift rapidly from one role to another. Understanding the relationships both within and outside an ecosystem is crucial for a platform strategy. In a platform context, S-D logic moves the focus on economic exchange and sequential value creation from a single organization to a broader actor-to-actor platform ecosystem network — comprising competitors, investors, suppliers, partners, regulators, and customers — in which a platform organization operates (platform ecosystem network-centric focus). Moreover, S-D logic emphasizes that tangible goods (products and services) are no longer the only object of exchange, but that there are also associated or stand-alone intangible offerings in which the extent of information content is high (data/information/knowledge-centric focus). S-D logic prompts considering platform value creation. It has influenced and proved the discussion of service ecosystems, which are defined as "a relatively self-contained, self-adjusting system of resource-integrating actors connected by shared institutional arrangements and mutual value creation through service exchange." (Vargo & Lusch, 2016, pp. 10–11) This service-ecosystems view highlights the role of context in value creation as a complex phenomenon emanating from several key constructs: service exchange, integration of resources, value co-creation, and value-in-context (Akaka, Vargo and Lusch, 2013; Akaka, Vargo, 2015).

To summarize, platform ecosystem perspective could be grounded in the following S-D logic (Vargo & Lusch, 2016, p. 18) derivative axioms:

- **Axiom 1:** Platform service is the fundamental basis of exchange.
- **Axiom 2:** Platform value is co-created by multiple platform ecosystem actors, always including the beneficiary.
- **Axiom 3:** All platform ecosystem social and economic actors are resource integrators.
- **Axiom 4:** Platform value is always uniquely and phenomenologically determined by the beneficiary.
- **Axiom 5:** Platform value co-creation is coordinated through actor-generated institutions and institutional arrangements.

With it's five axioms, a platform ecosystem increases availability and drives value creation in both global and local markets. Thus, value creation within global platform value chains is driven by the integration and application of resources within complex networks (e.g., organizations, industries) that interact across national and cultural borders.

At the same time, as it was mentioned by Parker and Van Alstyne (2015), boundaries of platform governance model distinguish a platform firm from the ecosystem in which it is embedded. Contributors to competing platforms, for example, are part of the ecosystem but need not abide by a focal platform's rules (Parker and Van Alstyne, 2015).

For global platforms, value creation within "global value chains" is driven by the integration and application of resources within complex platform networks (e.g., organizations, industries), which provides insight into understanding the dynamics of exchange relationships in multi-level markets (e.g., local, national, global). Importantly, this dynamic multi-faceted approach provides important insights into what underpins the complexity of the context that frames value creation in international markets (Akaka, Vargo and Lusch, 2013).

PLATFORM STRATEGY

Boudreau (2010) has stated that technology platforms can allow creation of the ecosystem around a new technology without giving up the ability to capture value. The high valuations of some platform technology owners have made platforms a focus of interest in the technology strategy. Owners of products or services with platform potential face several strategic questions:

- How can I transform my product into a platform?
- What kind of platform should I create?
- How can I transform my platform into a leader in the industry?

According to Gawer and Cusumano (2008), to have a platform potential, a product (or technology or service) "should perform at least one essential function . . . or solve an essential technological problem within an industry." Gawer and Cusumano (2008) suggest that firms that want to be platform leaders should follow two principal strategies. The first is the coring strategy: to transform a product into a platform. The second additional strategy is tipping: the process of building market momentum so that the industry tips toward wide adoption of the platform technology. Tactics can include penetration pricing and subsidies, building coalitions of customers and providing powerful incentives for complementors (Create). Gawer and Cusumano's strategic options are summarized in Table 1.

Table 1. Strategies to Become a Platform Leader: Coring and Tipping

Strategic Option	Technology Actions to Consider	Business Actions to Consider
Coring: How to create a new platform where none existed before	• Solve an essential "system" problem • Facilitate external companies' provision of add-ons • Keep intellectual property closed on innards of your technology • Create strong interdependencies between platform and complements	• Solve an essential business problem for many industry players • Create and preserve complementors' incentives to contribute and innovate • Protect your main source of revenue and profit • Maintain high switching costs to competing platforms
Tipping: How to win platform wars by building market momentum	• Develop unique, compelling features that are hard to imitate and that attract users • Tip across markets: absorb and bundle technical features from an adjacent market	• Provide more incentives for complementors than your competitors do • Rally competitors to form a coalition • Consider pricing or subsidy mechanisms that attract users to the problem

Source: Annabelle Gawer and Michael A. Cusumano, "How Companies Become Platform Leaders," *MIT Sloan Management Review* 49 (Winter 2008): 28–35. 2008 from MIT Sloan Management Review/Massachusetts Institute of Technology.

Network platforms differ from product platforms because stronger network effects, switching costs and multi-homing costs create greater pressure for market concentration (Parker and Van Alstyne, 2015). Firms that compete in such markets need clearer guidelines by which to set strategies to manage launching, openness and governance, as well as competition.

Definition 6. (Parker and Van Alstyne, 2015).

A *platform strategy* is the mobilization of a networked business platform to expand into and operate in a given market.

An effective platform strategy limits launch to a small community in order to generate strong, although bounded, network effects. eBay started as a market for Pez sweet dispensers (Evans, 2003). Diner's Club targeted restaurants and patrons on Manhattan Island (Evans and Schmalensee, 2010). Facebook launched exclusively among Harvard undergraduates before expanding to all '.edu' and then '.com' domains (Ellison et al., 2007). The idea is that a more sharply defined community will experience stronger network effects if a substantial fraction of the community adopts a particular platform product or service (Teece, 2012; Teixeira and Brown, 2016a). Once adoption takes place within that community, the platform can be opened to adjacent groups, pulling new users onto the platform (Teece, 2012; Teixeira and Brown, 2016b).

Platform strategy has the advantage of allowing the platform to build capacity in the following stages (Cusumano and Gawer, 2002; Eisenmann et al., 2006; Eisenmann et al., 2011; Teece, 2012; Parker and Van Alstyne, 2015):

- **Implementation**
- **Platform Design:** Combination of pricing and strategic decisions that affect each side; features to offer each side so as to reduce search, transaction costs or both
- **Mobilization and Monetization:** The order to get the sided on board; the platform owner faces a the "chicken-and-egg" problem
- **Platform Control:** Who gets access to what side, and what each side is allowed to do on the platform
- **Envelopment:** Launching into an adjacent market from an established platform is the process of 'platform envelopment' (Eisenmann, Parker, Van Alstyne, 2011; Teece, 2012).
- Others

Platform strategy becomes extremely complex as companies consider dynamic interactions of a multi-layered business ecosystem (Teece, 2012).

PLATFORM LIFE CYCLE

A platform ecosystem usually goes through periods of competitive strength and weakness as external circumstances change, and these dynamics also need to be managed by the platform owner. Teece (2017a,b) described a platform ecosystem lifecycle as consisting of four phases: Birth, Expansion, Leadership, and Self-Renewal (see Figure 3). This model provides a useful framework for thinking about how platforms and mentioned dynamic capabilities interact.

Figure 3. Platform dynamic capabilities over a platform lifecycle (adopted from Teece, 2017a.b)

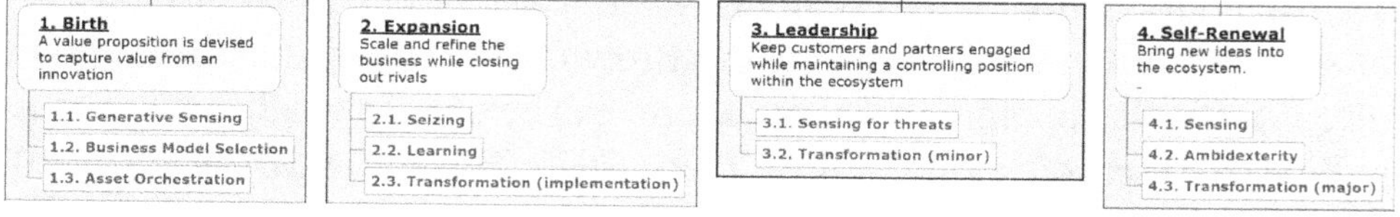

In the *birth phase*, once an opportunity has been sensed, the next step is to design a business model. Enterprise's ability to develop and refine platform business models is the core microfoundation of platform dynamic capabilities. A successful platform business model solves a problem for an identifiable group of potential customers who can support a high enough price to cover all costs and leave a satisfactory profit.

In reality platforms use nuanced combinations of legal, technological, informational and other instruments (including price-setting) to implement desired outcomes. The following taxonomy (Figure 4) categorizes these strategies (Coles and Edelman, 2011).

Figure 4. Practical platform strategies of preventing platform disintermediation (adopted from Coles and Edelman, 2011)

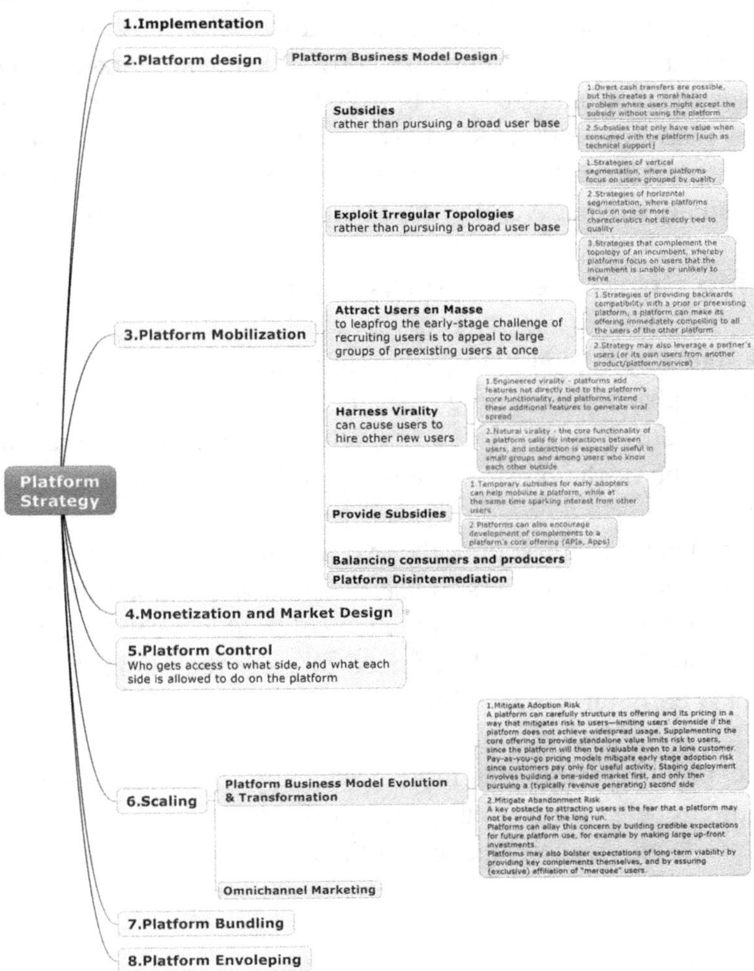

PLATFORM MOBILIZATION

Network effects are particularly predominant in platform businesses. The central problem challenging platforms subject to network effects is how to achieve enough user adoption to reach the critical mass (Evans and Schmalensee, 2010). That's why platforms face a unique set of barriers in attracting an initial group of users. As user's value from platform participation depends on the number of other users,platform owner faces a chicken-and-egg problem: how can a platform ever achieve the "critical mass" of users if users refuse to join unless there is already a critical mass? How to reach the critical mass of side users affiliated to the platform? Companies have used a number of strategies to overcome this issue. Several authors consider strategies for attracting an initial group of users, including subsidy platforms with substantial resources, exploiting irregular topologies, harnessing virility and creating "stand-alone" value (Parker and Van Alstyne, 2005; Eisenmann, Parker, Van Alstyne, 2006; Eisenmann, 2007; Eisenmann, Parker, Van Alstyne, 2011; Coles and Edelman, 2011; Hagiu, 2014; Teixeira and Brown, 2016a,b).

Subsidy Platforms with substantial resources can entice users via subsidy to join the platform. Subsidies can be temporary penetration prices or permanent discounts and can take several forms (Parker and Van Alstyne, 2005).

Balancing Consumers and Producers

As it was stated by P. Hu (2015), acquiring platform consumers is the focus early on, but it also becomes progressively harder to grow platform producers at a suitable rate. For two-sided marketplaces balancing consumers and producers is one of the most challenging problems in the growth stages of a company.

In order to get initial traction, the two- or multi-sided marketplaces need both consumers and producers to use the platform. Onboarding producers early on is critical because of the frequency of producer's transactions. There are two reasons why marketplaces need to grow the number of producers more quickly early on (Hu, 2015):

1. Since producers transact more frequently, they are instrumental in creating repeat business early on (see Table 2).
2. Idle producers are less dangerous than idle consumers.

Table 2. (Hu, 2015).

Platform	Transaction per Producer / Transaction per Consumer
AirBnB	70:1
Uber	50:1
Ebay	5:1

Optimal growth in the number of producers and consumers is determined by customer lifetime value (Hu, 2015). Optimizing the growth in number of producers and consumers depends on many factors, including timing and location peculiarities. However, the approximate numbers can be assumed by looking at retention rates and transaction frequencies.

In a marketplace that fully satisfies producers and consumers, the following equation holds (Hu, 2015):

(number consumers) * (buyer lifetime value if all transactions are satisfied) = (number of producers) * (seller lifetime value if all transactions are satisfied).

Therefore,

(number of consumers) / (number of producers) = (seller lifetime value) / (buyer lifetime value)

By the definition of lifetime value (discount rates are ignored for this analysis since it remains consistent for consumers and producers),

(Lifetime value) = (Gross margin per transaction) * (monthly frequency) / (monthly churn)

The gross margin per transaction is the same for consumers and producers from the point of view of the marketplace. Therefore,

(number of consumers) / (number of producers) = (seller frequency) / (buyer frequency) * (buyer churn) / (seller churn)

Churn also depends on frequency, though the relationship differs by marketplace. In general, though, the less frequent a party performs a transaction, the higher the likelihood of churn (Hu, 2015).

As we can see from Table 1, the frequency of transactions for an Uber driver is much higher than the frequency of transactions for an Uber rider. Relatedly, the frequency of transactions for an Airbnb host is much higher than for an Airbnb guest. Looking forward, these companies will have to grow their consumer (buyer) base much more quickly than their producer (seller) base. As these companies determine their steady-state churn rates on both sides over time, the data could suggest that the growth needs to be more lopsided than before.

We could find an important contrast in the traditional Ebay two-sided marketplace platform (Table 2). Ebay producers do transact more frequently than consumers, but the ratio is an order of magnitude smaller compared to Uber and Airbnb. Growth in these platform marketplaces cannot be measured the same way as it was measured for Ebay.

Two- or multi-sided marketplaces face many challenges, and balancing buyer and seller growth is only one of them.

Platform Disintermediation

Other challenges include how to overcome the chicken-and-egg problem when launching as well as how to keep transactions on the platform after a buyer and a seller have already found each other. This latter problem is known as two-sided platform disintermediation: users may rely on the marketplace to find each other but then perform related future transactions — or even the current transaction — without the platform's involvement and without paying any fees the platform may charge (Edelman and Hu, 2016).

Disintermediation can be caused by the following two distinct contexts for connected but distinct reasons (Edelman and Hu, 2016):

- **Disintermediation for the Current Transaction:** When a marketplace's matchmaking function directly connects two parties, they may choose to depart the platform to complete the transaction, including contracting and payment. In this case, the platform is not capable to charge fees for transaction, and the platform usually does not even receive consistent information about what happened. The larger the value of the current transaction, the greater the parties' incentive to circumvent the marketplace and its fees.

- **Disintermediation for Subsequent Transactions:** In this case parties may choose to complete the current transaction through the platform, but later run additional transactions with each other outside of the platform. This approach is particularly likely if the nature of the service invites a consumer to continue with a single service provider.

Adopted framework (see Figure 5) could be used for assessing which strategies of preventing disintermediation are most relevant (green color) to alternative types of disintermediation.

Figure 5. Strategies of preventing platform disintermediation (adopted from Edelman and Hu, 2016)

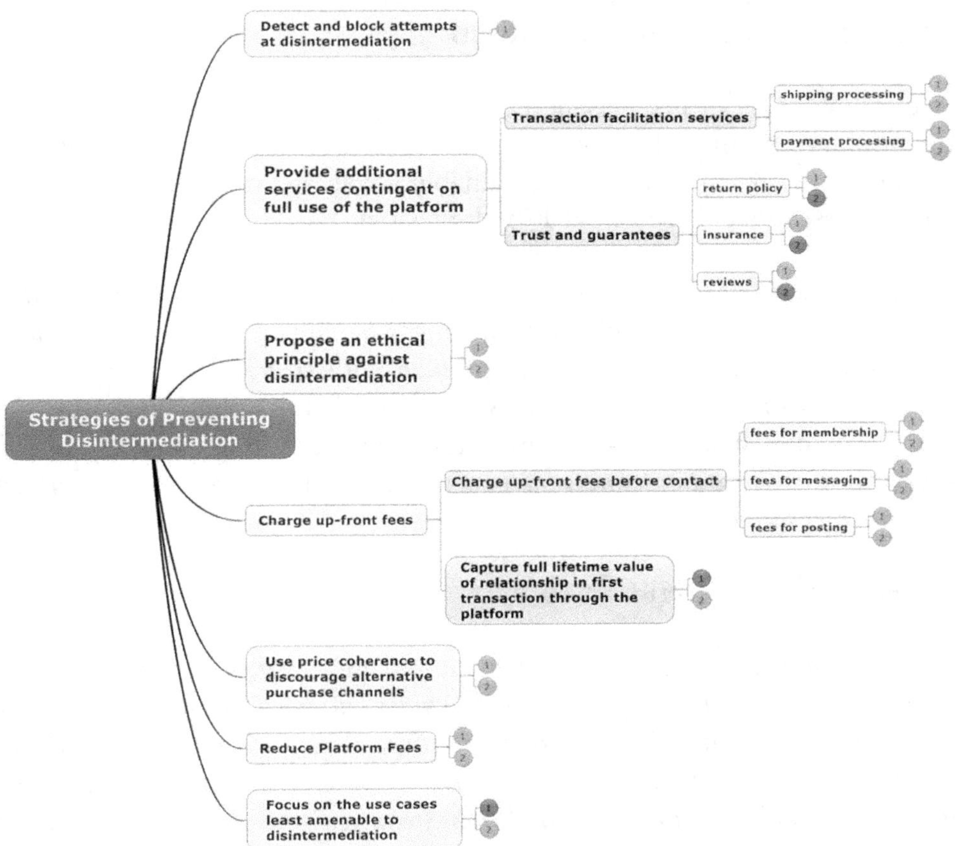

Entry Barriers

There are several ways for MSPs to overcome entry barriers, such as strong network effects or high switching costs in multi-sided markets:

- To provide technological features that are far superior to those of the dominant incumbent (Evans and Schmalensee, 2002).
- To propose platform envelopment as another path of entry (Eisenmann et al., 2011). Envelopment happens when a new platform powers common elements (such as similar product functionalities or overlapping user bases) between different markets through bundling or co-branding, to enter an adjoining market.

Platform Strategy, Dynamic Capabilities, and Platform Business Models

According to Casadesus-Masanell and Ricart (2011:100): "Strategy has been the primary building block of competitiveness over the past three decades, but in the future, the quest for sustainable advantage may well begin with the business model". Teece (2017b) argues that

It is perhaps more accurate to say that unique capabilities are the primary building block of firm-level competitiveness because they enable business model design, which is deeply intertwined with strategy. In many cases, corporate strategy dictates business model design. At times, however, the arrival of a new general-purpose technology (e.g., the Internet) opens opportunities for radically new business models to which corporate strategy must then respond.

Teece (2017b) describes:

That dynamic capabilities ... include the sensing, seizing, and transforming needed to design and implement a business model. They can enable an enterprise to upgrade its ordinary capabilities and direct these, and the capabilities of partners, toward high-payoff endeavors. This requires developing and coordinating, or "orchestrating," the firm's (and partner firms') resources to address and even shape changes in the marketplace, or the business environment more generally.

Figure 6. Strategy, Dynamic Capabilities and Platform Business Models framework

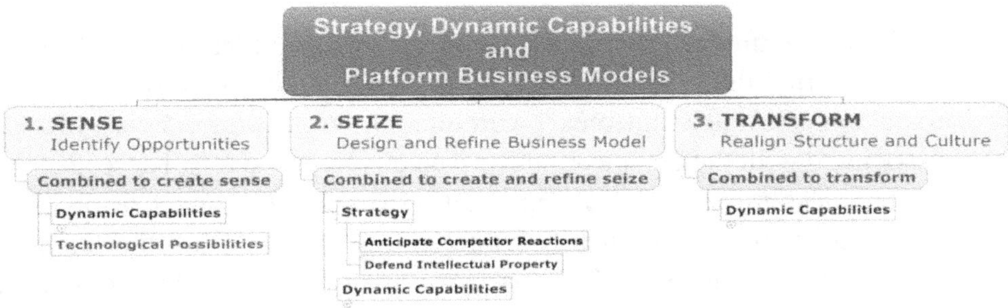

The framework (Figure 6) based on dynamic capabilities framework (Teece, 2017b) reveals platform strategy, dynamic capabilities and business model interdependence. Dynamic capabilities and platform strategy conjoint to create and refine a platform business model that guides organizational platform transformation.

PLATFORM BUSINESS MODELS

Platform business models always should be related with *platform strategy* (Eisenmann et al., 2006; Eisenmann et al., 2011).

One key choice for a platform strategy is to select the right platform business model. The different elements of an organization's strategy and it's business model have to be aligned and coherent (Teece, 2017b).

Business Models

The business model concept became prevalent with the beginning of the Internet in the mid 1990s (Timmers, 1998; Zott, Amit and Massa, 2011). Since then at a general level, business model has been referred to as a conceptual tool or model (Osterwalder, 2004; Osterwalder, Pigneur and Tucci, 2005; Osterwalder, Pigneur, 2010), a framework (Afuah, 2004), a pattern (Brousseau and Penard, 2006; Osterwalder and Pigneur, 2010). Although there are many definitions of business models (Osterwalder and Pigneur, 2010; El Sawy and Pereira, 2013) there is no one accepted definition of it (Zott, Amit, and Massa, 2011).

Some definitions of business models are, as follows:

1. "A business model articulates the logic and provides data and other evidence that demonstrates how a business creates and delivers value to customers. It also outlines the architecture of revenues, costs, profits associated with the business enterprise delivering value" (Teece, 2000).
2. "A business model is a conceptional tool containing a set of objects, concepts and their relationships with the objective to express the business logic of a specific firm. Therefore, we must consider which concepts and relationships allow a simplified description and representation of what value is provided to customers, how this is done and with which financial consequences" (Ostenwalder and Pigneur, 2010).
3. "Business model as the description of the articulation between different business model components or building blocks to produce a proposition that can generate value for consumers and thus for the organization" (Lecocq and Demil, 2010).
4. "In general, the business model can be defined as a unit of analysis to describe how business is transacted. More specifically, the business model is often depicted as an overarching concept that takes notice of the different components a business is constituted of and puts them together as a whole ...In other words, business models describe how the magic of a business works based on its individual bits and pieces" (Gassmann et al. 2014 b).
5. "A business model consists of two essential elements — the value proposition and the operating model — each of which has three sub elements. The value proposition answers the question: what are we offerings to whom? It reflects explicit choices along the following dimensions:
    a. **Target Segment(s):** Which customers do we choose to serve? Which of their needs do we seek to address?
    b. **Product or Service Offering:** What are we offering the customers to satisfy their needs?
    c. **Revenue Model:** How are we compensated for our offering?

The operating model answers the question: how do we profitably deliver the offering? It captures the choices the business needs to make in the following critical areas:

a. **Value Chain:** How are we configured to deliver on customer demand? What do we do in-house? What do we outsource?
b. **Cost Model:** How do we configure our assets and costs to deliver on our value proposition profitably?
c. **Organization:** How do we deploy and develop our people to sustain and enhance our competitive advantage?" (Deimler, Lesser, Rhodes, Sinha, 2013).

Business Model Elements

Business model is a useful lens for understanding a company's fundamental value creation logic. To achieve a common understanding of the business model concept, several authors have identified elements belonging to a business model (e.g., Hedman and Kalling, 2003; Johnson, 2010; Osterwalder and Pigneur, 2010). Today the interpretation of business models as formal conceptual representations/descriptions becomes predominate (Osterwalder, 2004; Gassmann et al., 2014a,b). Osterwalder (2004) defined a business model as "a conceptual tool that contains a set of elements and their relationships and allows expressing the business logic of a specific firm. It is a description

Figure 7. The taxonomy of business model building blocks (adopted from Osterwalder and Pigneur, 2010)

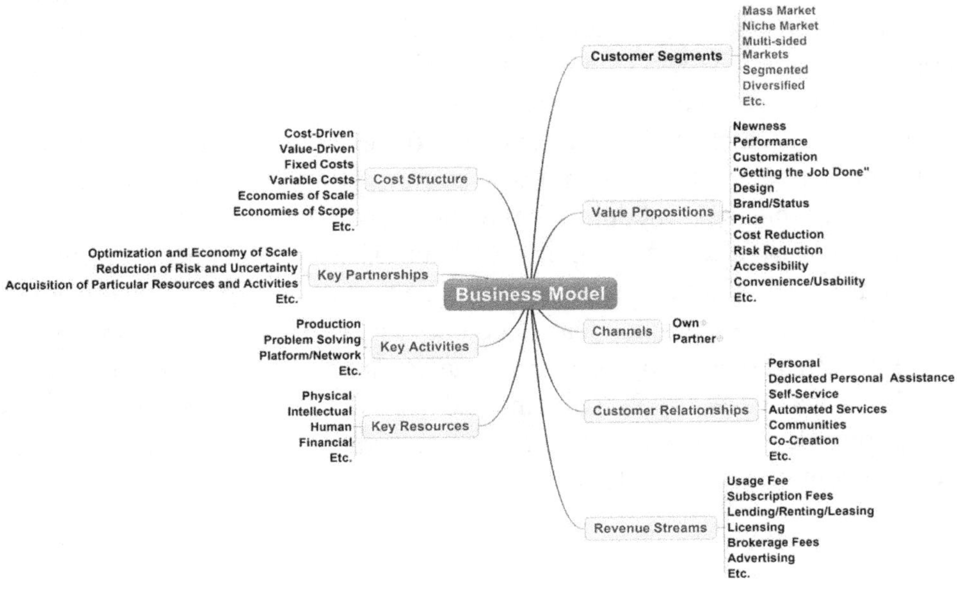

Figure 8. Business model building blocks with meta-components (adopted from Günzel and Holm, 2013; Gassmann et al., 2014a,b)

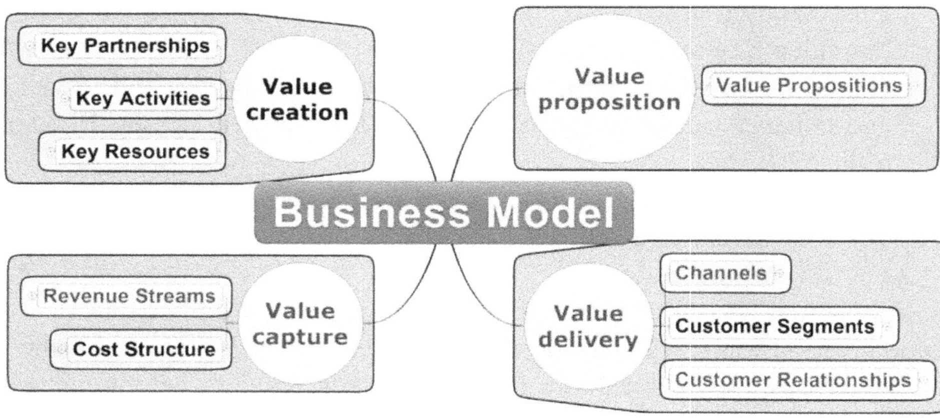

of the value a company offers to one or several segments of customers and the architecture of the firm and its network of partners for creating, marketing and delivering this value and relationship capital, to generate profitable and sustainable revenue stream".

Osterwalder (2004) has provided a business model ontology, which is a conceptualization and formalization of the elements, relationships, vocabulary, and semantics of a e-business model. This ontology is structured into several levels of decomposition with increasing depth and complexity. To abridge a complicated ontology, Osterwalder and Pigneur (2010) have developed a popular Business Model Canvas - simplified taxonomy of 9 main business model building blocks (see Figure 7, 8).

Several researchers have attempted to describe and classify generic e-business models and typologies. Timmers (1998) distinguishes among 11 generic e-business models, from e-shops and e-procurement to trust and other third-party services. Rappa (2001) classifies companies according to the nature of their value propositions and their modes of generating revenues. Since then business models have become a main stream and connected to more generic research approaches and design perspectives (Bouwman et al., 2012). As was shown by Zott & Amit (2001), a business model characterizes an opportunity for value creation through four value drivers: novelty, lock-in, complementarities, and efficiency.

Business Model Innovation

Teece (2017b) showed that ¨...Business models are enabled by dynamic capabilities in the sense that a dynamically capable organization will be able to rapidly implement, test, and refine new and revised business models. Successful implementation draws on management's architectural design, asset orchestration, and learning functions, which are core dynamic capabilities. At the same time, dynamic capabilities depend in part on the organizational flexibility allowed or denied by business model choices such as whether to outsource the manufacture of a new product or build a factory¨.

The researcher argues about the importance of dynamic capabilities for business model innovation and the importance of organizational design for both business models and dynamic capabilities.

Considering static business models we use the business model concept as a means to take a snapshot of an organization. However, the current and future states of a company's business model are connected by it's dynamic capabilities – it's abilities to reconfigure it's assets (Teece, 2017b). The idea of business-model innovation involves making simultaneous, coordinated and internally consistent changes to multiple aspects of the business model to reignite growth, combat disruptions or access new markets.

Business model innovation (BMI) is defined by Zott and Amit (2015) as "the design and implementation of an activity system that is new to the focal firm or new to the product–market space in which the focal firm competes." Researchers argue that it matters to managers, entrepreneurs, and academic researchers for several reasons (Zott and Amit, 2015):

1. It represents an often underused source of value. Business model represents an opportunity for value creation through four value drivers: novelty, lock-in, complementarities, and efficiency.
2. Competitors might find it more difficult to imitate or replicate an entire novel activity system than a single novel product or process. As it is relatively easy to undermine and erode the returns of product or process innovation, innovation at the level of business modelling could translate more readily into sustainable competitive advantage.
3. Because BMI can be such a potentially powerful competitive tool, managers should be attuned to the possibility of competitors' efforts in this area. Competitive threats often come from the outside boundary of the traditional industry, especially in platform markets (Uber, Google etc.).

Figure 9. Business model innovation process framework (adopted from Frankenberger et al., 2013)

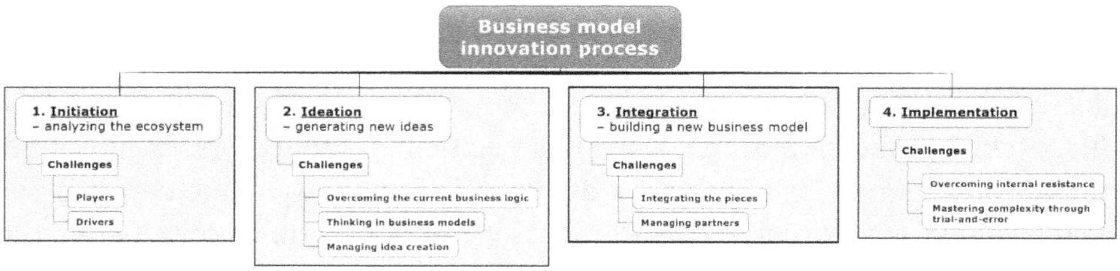

Frankenberger et al. (2013) developed an integrative business model innovation process framework which includes the structure and challenges associated with business model innovation (see Figure 9).

Two- or Multi-Sided Platforms

The main types of platform business models are shown on Figure 10. Selecting a platform business model could reformulate more precisely what type of business the owner of the platform technology wants to become.

Internet intermediaries or digital intermediaries can be defined as organizations that bring together or facilitate transactions between third parties on the Internet or other networks. Currently on the digital business market, the dominant definition of an intermediary business model refers to a platform intermediary, where two or more agents interact through an intermediary two or multi-sided platform (Armstrong, 2006) that form two-sided and multi-sided markets.

So what makes multi-sided platforms special and different from regular firms and other intermediaries? The most common approach to date has focused

Figure 10. Platform business models

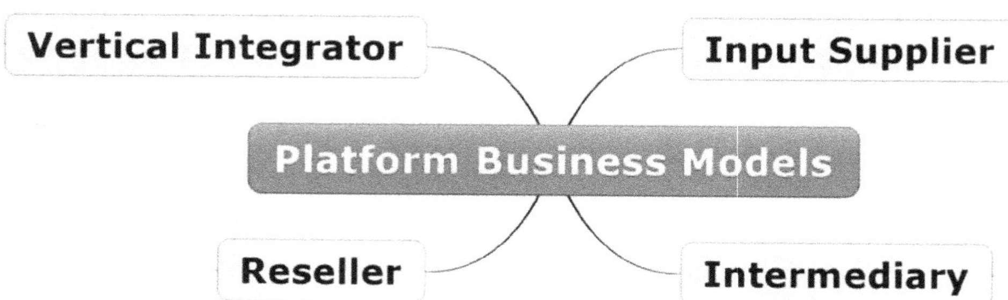

on the presence of important cross-group or indirect network effects between two or more customer groups participating on the platform (Armstrong, 2006). Rochet and Tirole (2003) suggested an alternative definition, which emphases on whether the structure of prices set by the platform is non-neutral (i.e., whether the allocation of fees across the two sides matters for the total volume of transactions). Both approaches have limitations (see Rysman, 2009). For instance, suppliers of supermarkets and discount superstores that take on inventory risk care about the number of consumers visiting the stores (and vice-versa), so that indirect network effects exist.

Furthermore, if supermarkets and discount superstores pay their suppliers more and increase their prices to consumers, this change is unlikely to be neutral for their sales. Thus, such stores along with most other retailers appear to be MSPs according to the definitions above, yet most economists would agree that they are not. Researchers use the terms "two-sided market", "two-sided platform" or "multi-sided platform" to refer to the mediating role of service platforms between two or more groups of agents (Evans et al., 2006, 2011, 2014; Rochet and Tirole, 2003, 2004). According to the definition (Evans et al, 2006; Hagiu, and Wright, 2015b) a multi-sided market exists, when at any point in time there are:

- Two or more distinct groups of customers;
- The value obtained by one kind of customers increases with the number of the other kind of customers; and
- An intermediary is necessary for internalizing the externalities created by one group for the other group.

Such intermediaries are known as Two- or Multi-Sided Platforms. Hagiu and Wright (2015b) proposed a definition that provides a more precise notion of Multi-Sided Platform by requiring that they enable direct interactions between the multiple customer types that are affiliated to them:

Defenition 6. (Hagiu and Wright, 2015b)

A multi-sided business platform (henceforth, MSP) is an organization that creates value primarily by enabling direct interactions between two (or more) distinct types of affiliated customers.

At the most fundamental level, MSPs have two key features (Hagiu and Wright, 2015b):

- They enable direct interactions between two or more distinct sides
- Each side is affiliated with the platform

By "direct interaction" Hagiu and Wright (2015b) mean that "the two or more distinct sides retain control over the key terms of the interaction, as opposed to the intermediary taking control of those terms." Where the interaction involves trading, the key terms of the interaction could be the pricing, bundling, marketing and delivery of the goods or services traded, the ability to determine the nature and quality of services offered, the terms and conditions, etc."

By "affiliation" Hagiu and Wright (2015b) mean that "users on each side consciously make platform-specific investments that are necessary in order for them to be able to directly interact with each other. The investment could be a fixed access fee (e.g., buying a videogame console), expenditure of resources (e.g., spending time and money on learning how to develop applications using the iPhone's APIs), or an opportunity cost (e.g., driving to a shopping mall, joining a loyalty program)."

Each of these features helps differentiate MSPs from related but distinct business models. Direct interactions between multiple sides set MSPs apart from resellers and fully vertically integrated firms. Affiliation by all relevant customer types (sides) helps distinguish MSPs from input suppliers which are not "approved" by all sides (Hagiu and Wright, 2015b). This difference holds even when the inputs are "needed", which the management literature calls "platforms" (Gawer and Cusumano, 2008). Hagiu and Wright (2015b) state that affiliation by multiple sides is necessary for MSPs to create cross-group network effects – the key defining property of MSPs put forward in much of the existing literature.

Figure 11 represents in a simple way the key distinctions between MSP and input suppliers on the one hand and between MSP and re-sellers on the other hand (Hagiu and Wright, 2015a). An intermediary can choose between functioning as a marketplace, on which suppliers sell their products directly to buyers, or as a reseller, purchasing products from suppliers and selling them to buyers (Hagiu and Wright, 2015a, 2015b). In the model proposed by Hagiu and Wright, this choice comes down to whether residual control rights over non-contractible decision variables are better held by suppliers (the marketplace mode) or by the intermediary (the reseller mode). They focus on a single non-contractible decision variable – the level of demand – enhancing marketing activities. The reseller is better able to coordinate these activities across products, whereas the marketplace mode benefits from allowing independent suppliers to tailor their marketing activities to local information about buyer demand. In the benchmark setting, the marketplace mode is preferred if and only if the variance of local information exceeds the squared value of spill overs from marketing activities across products. Amazon, for example, acts as an online retailer for most products, choosing which products to sell and what prices to ask. Suppliers and customers do not interact directly. But Amazon also provides a platform over which people can interact directly to buy and sell used books, and Amazon collects a commission. The company has reportedly found that the margins from providing a platform are greater than those from operating as a reseller. Apple also uses both models. For music and movies sold over iTunes, Apple acts as a reseller. But Apple's App Store provides a platform for direct interaction between buyers and sellers of application software. Hagiu and Wright (2013) discuss the factors companies should consider in selecting a particular platform model. They explore several generalizations, showing how the benchmark trade-off is modified in different settings.

In platform industries, interaction follows a triangular relationship (Eisenmann et al., 2011), as the parties first affiliate with the platform and then connect or trade using platform resources. For example, on Airbnb, renters and hosts transact with each other but they use the platform to match, to perform searches, to enter into contractual agreements, to transfer payment, to acquire insurance and manage reputations to facilitate future transactions (Parker and Van Alstyne, 2015).

In two-sided and many-sided markets the value creation and appropriation logic is different, so the linear and transitive value chain and value creation thinking is outdated. Often this duality of both sides representing cost and revenue is neglected and one side is treated as a profit on, while the other either as a loss or as a financially neutral (Hagiu and Wright, 2015b).

MSP users are permanent members of one distinct group – a "side" – which transacts with the second group, e.g.:

- Job seekers + recruiters;
- Card holders + merchants.

Figure 11. Difference between input suppliers, MSP's and re-sellers caption (Hagiu and Wright, 2015b)

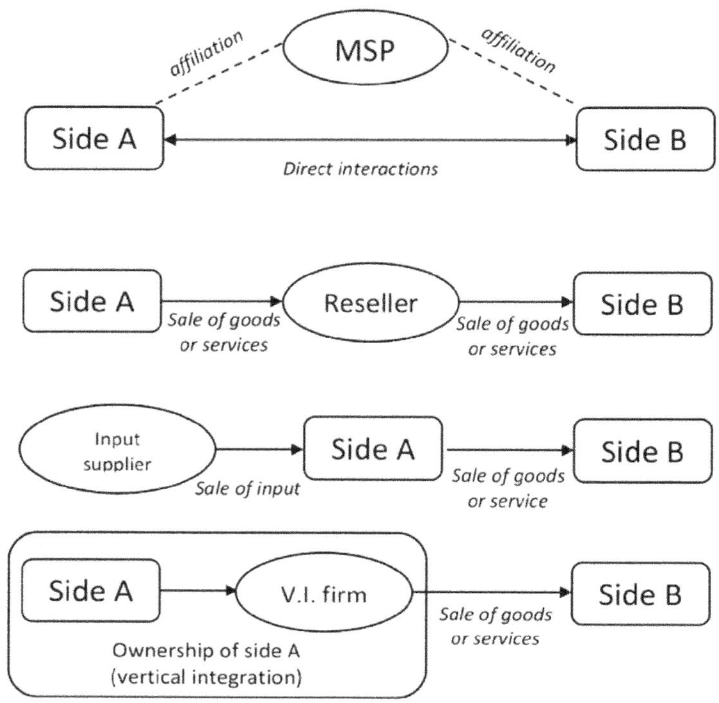

MSPs guide technological innovation trajectories and stimulate innovation on complements. Network users access a common platform that facilitates their interactions.

Examples of MSPs include:

- *Google's search engine* brokers the relationship between online advertisers and Internet users.
- *Amazon, eBay* and other e-commerce sites connect buyers and sellers.
- *American Express, PayPal, Visa* and other payment systems bring together merchants (online or offline) and consumers.
- *Microsoft's Windows* coordinates PC manufactures, third-party developers of software applications and end-users.
- *Social web (Facebook, Twitter, Google+)* brings together users, advertisers and third-party developers of web applications.
- *Mobile service platforms* (Apple with App Store) connect together users, advertisers and third-party developers of mobile applications.

- *Videogame consoles* such as Nintendo's Wii, Sony's PlayStation 3 and Microsoft's Xbox 360 connect users with third-party game developers.
- *Cloud computing platforms* (AWS, Microsoft).
- Etc.

MSP comprise mostly all possible types of transactions: B2C, B2B, C2C/P2P, B2G, C2G, etc. MSPs are not just digital industries, but also:

- Financial services, e.g., ATMs, credit cards, stock exchanges
- Transportation, e.g., package delivery, airlines, travel agents, reservation systems, fuel cell-powered cars
- Retail, e.g., shopping centers, bar codes/RFID
- Energy, e.g., grid + appliances, energy trading
- Real estate, e.g., home buying
- Health care, e.g., HMOs
- Enterprise administration, e.g., recruiting, B2B procurement

Multi-Sided Platform BMP

MSP business models are characterized by high evolutionary dynamics: the number of sides, products and/or services is steadily increasing. For example, Amazon started in 1995 as an online book retailer, then in 1999 it added e-auctions and zShops, in 2000 – e-marketplace, in 2001 – partnerships and logistic services, in 2006 – Amazon (Cloud) Web Services, in 2007 – eBooks/Kindle.

A clear and precise description and structuring of information of [multi-sided] platform business models are prerequisites for a common understanding of the digital platforms.

Osterwalder and Pigneur (2010) proposed Business Model Patterns (henceforth, BMPs) – business models with similar characteristics, similar arrangements of business model building blocks or similar behaviors. BMPs help to understand business model dynamics and serve as a source of inspiration for creating e-commerce business models (Gassmann et al., 2014a,b). Business Model Patterns concept is another way of structuring digital value drivers (Gassmann et al., 2014a,b). Business model patterns highlight a number of specific design ideas and detail them by using companies having successfully implemented them as examples.

Following Osterwalder (2004) we provide a multi-sided platform business model pattern ontology, which is a conceptualization and formalization of the elements, relationships, vocabulary and semantics of a digital intermediaries MSP (Yablonsky, 2013). It is structured into several levels of decomposition with increasing depth and complexity. Ontologies, taxonomies and other types of controlled vocabularies are the preferred means to achieve such a common understanding by specifying the terms of the e-intermediary's domain, disambiguating them from each other, controlling synonyms and structuring the domain via term relationships (Poli et al., 2010; Yablonsky, 2014a). The total number of all ontology features is too complex to be represented here in its entirety, but an example of taxonomy is provided in order to demonstrate both the process of classification and the intermediate result. A taxonomy comprises a hierarchy of concepts linked by a transitive subsumption relation (often called isA or subClassOf) whereby each instance of a class can be inferred to be an instance of all parent classes (Poli et al., 2010). Taxonomies are strict hierarchies: each class has at most one parent. Therefore, MSP BMP framework has a distinct structure of taxonomy and is shown on the Figure 12.

MSP BMP creates *Value* by facilitating interactions between different sides of the platform. The MSP BMP *Value Preposition* usually creates value in the following main areas (Hagiu, 2006; Osterwalder and Pigneur, 2010; Evans, 2013; Hagiu and Wright, 2015 a,b):

- Mobilizing user groups (Customer Segments) and balancing interests;
- Matchmaking between Customer Segments;
- Reducing costs by channeling transaction through the platform;
- Strategic decisions in platform evolution, multi-homing costs, scaling and liquidity.

MSP grows *value* to such an extent that it attracts more users. MSP's value to a user depends on the number of other platform users and defines Willingness-to-Pay (henceforth, WTP) for *platform participation*. WTP for *platform affiliation* is a cap on platform fees (Eisenmann et al., 2006).

The key to MSPs is to understand the network effects which occur when the value of a product depends on the number of other users (Shapiro and Varian, 1999). A *network effect* (also called *network externality*) is the effect that one user of a platform good or service has on the value of that product to other people. When a network effect is present, the value of a product or service is dependent on the number of others using it. Network effect can be *negative* (e.g., due to congestion) or *positive*.

Figure 12. MSP BMP Framework (developed from Osterwalder and Pigneur, 2010)

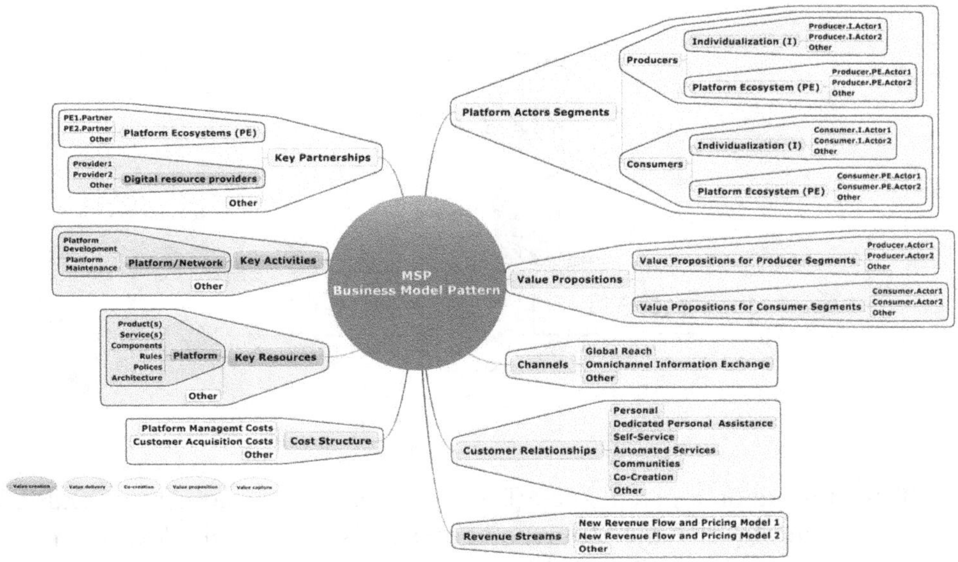

A two-sided platform has four network effects (Figure 13):

- A *same-side* effect for each side, i.e., preference regarding the number of other users on the own side.
- A *cross-side* effect in each direction, i.e., preference regarding the number of users on the other side.
- Each effect can be positive or negative.

An example of one direction cross-side network effect are the advertisers on Google who are attracted to the large user base of online Google's searchers. Google's searchers, nevertheless, are not attracted by the number of advertisers who have access to the Google search engine. Both direction cross-side effects are current in card payment systems where the presence of the cardholders is dependent on the presence of merchants and vice versa.

With positive network effects a larger MSP user base is appealing and the platform network is growing. MSP network growth increases the willingness to pay (WTP) of the prospects who had not yet acquired the MSP. Strong and positive network effects by increasing returns can bring some MSP to Winner-Take-All (WTA) position, when one MSP serves the mature multi-sided market. There are many examples of WTA MSPs: Windows, e-auction eBay, Adobe file format PDF, search platform Google, etc.

Figure 13. Two-sided platform network effects (adopted from Eisenmann, 2007; Yablonsky, 2013, 2016c)

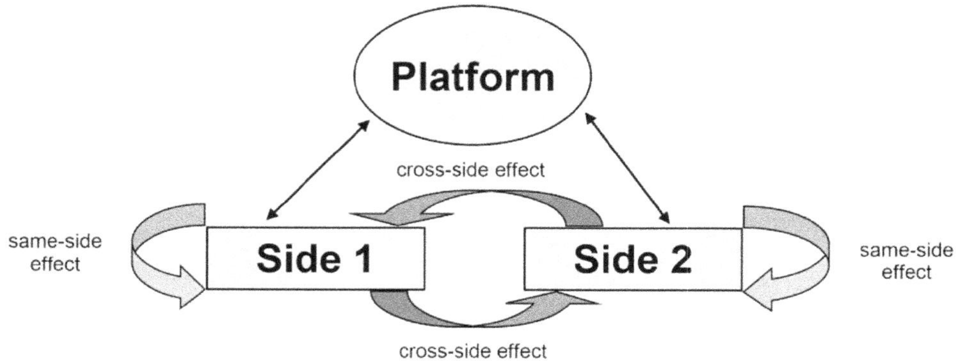

If most users on a given side affiliate with a single platform, it is called *mono-homing* (Eisenmann, 2007), otherwise it is called *multi-homing* (most users on a given side affiliate with multiple platforms).

Due to network effects, platform intermediaries often enjoy increasing returns to scale and their industries have room for only a few number of players (Parker and Van Alstyne, 2005). In many MSPs, network effects are so strong that a single platform prevails. With increasing shifts and changes of business landscapes, companies in both developed and emerging economies are challenged by how to manage and to innovate through MSPs (Eisenmann, Parker, Van Alstyne, 2006). When winners take most, little is left for losers, as evidenced by the spectacular success of some platform providers.

Cross-side network effects are such an important factor in determining which platforms come to dominate an industry that some researches define a platform as something that exhibits those effects. Payment platforms such as Visa and PayPal, online dating platforms such as Match.com, social media platforms such as Facebook, trading platforms such as eBay, Amazon, Baidu and gaming platforms such as Sony PlayStation have all become powerful in their industries because of indirect network effects. Although not all technology platforms exhibit powerful network effects (e.g., the Boeing airframe), those that do have a better chance of growing to a position of dominance in their industries.

A multi-sided market is more likely to be served by a single MSP when

- MSP is a natural monopoly or
- Multi-homing costs are high and network effects are positive and strong, while the demand for differentiated features is weak or the dominant MSP can offer such features selectively to users willing to pay premium (Eisenmann, 2007).

Eisenmann (2007) defines that network effects are demand-related scale economies because they influence the revenue side of platform provider's profit equation through their user adoption and WTP. Enterprises enjoy scale economies when their profit margins progress with expanding sales volume.

MSP BMP has two or more *Platform Customer Segments*, each of which has its own Value Preposition and associated Revenue Stream. Furthermore in some business models one Customer Segment cannot exist without another. Sometimes one or more Customer Segments may enjoy free offers or reduced prices subsidized by revenues from other Customer Segments. Choosing which segment to subsidize can be a crucial strategic pricing decision that determines the success of MSP BM. MSPs are of value to one group of customers only if the other groups of customers are also present.

The main *Platform Costs* incurred under MSP BMP relate to maintaining and developing the platform. Platform Costs can be divided into

- Homing costs, related to the adoption, operation, and/or other costs encountered due to platform affiliation (Armstrong, 2006),
- Switching costs, paid by consumers for switching from one platform to another (Shapiro and Varian, 1999).

Platform Business Model Design Methodology

Digital transformation challenges all business leaders with the need to reimagine their business in terms of becoming digital. However, currently the tools for this endeavor are in their early stage.

We use Business Model Pattern (BMP) concept developed by (Osterwalder and Pigneur, 2010; Gassmann et al., 2014 a,b; Remane et al., 2017) as a way of structuring platform digital value drivers and platform BMP. Several BMP research methodologies have been proposed during the last decade.

Gassmann et al. (2014 a, b) employ a conceptualization that consists of four central dimensions: the Who, the What, the How, and the Why. They used a two-step approach to analyze the basic patterns of business models. In phase 1, 250 business models that had been applied in different industries within the last 25 years were analyzed. As a result 55 patterns of business models that served as the basis for new business models in the past were identified. In phase 2, a new construction methodology for new business models design was developed, including the BMI Navigator that structures the process of innovation of a company's business model.

Osterwalder and Pigneur (2010), Remane et al. (2017) use the 9 block structured BMP. Remane et al. (2017) state that even when applying the most comprehensive collection with 55 business model patterns from Gassmann et al. (2014a,b), one misses more than two-thirds of the available patterns. This diversity in research influenced researchers to review 356 BMP mentioned in the 22 original sources. The result of research filtering resulted in creation of a database with 182 business model patterns. This database contains all generic and e-business-specific business models. The patterns in the database are classified along the 12 dimensions of the taxonomy of business model patterns. Each pattern is assigned to exactly one characteristic for each of the 12 dimensions. The business model pattern database reduces the decision making process, as it helps to distinguish the relevant set of patterns for a specific purpose depending on the specific situation of the innovating firm.

Therefore, one cannot fully agree with the opinion expressed by Blaschke et al. (2017a) that business model patterns and digital value drivers differ in one central aspect: while business model patterns address the entire business model, digital value drivers focus on particular business model components and digital key elements. We believe that granularity of BMP concept is proved by several successful taxonomies of real business model patterns.

We provide platform BMP namely MSP BMP as a basic conceptual framework and knowledge management tool for describing, analyzing and interpreting non-price instruments used by digital platforms, especially by platform intermediaries, as "private regulators" who regulate access to and interactions around different multi-sided markets (Evans et al., 2011).

Using MSP BMP Framework (see Figure 12) we adopt the 12 dimensions of the business model pattern taxonomy (Remane et al., 2017) for defining 14 dimensions of the MSP business model pattern taxonomy (see Figure 14).

Using the MSP BMP is the basis of digital intermediaries analysis and decision making. For each industry we update this pattern, converting it into a MSP BMP for a specific type of industry platform (search MSP BMP, e-retailor MSP BMP etc.). Finally, for a platform entity – a given MSP company industry/type (Google, Facebook, Amazon, startup, etc.) – industry MSP BMP is transferred to a company's MSP BM:

Figure 14. Platform BMP dimensions

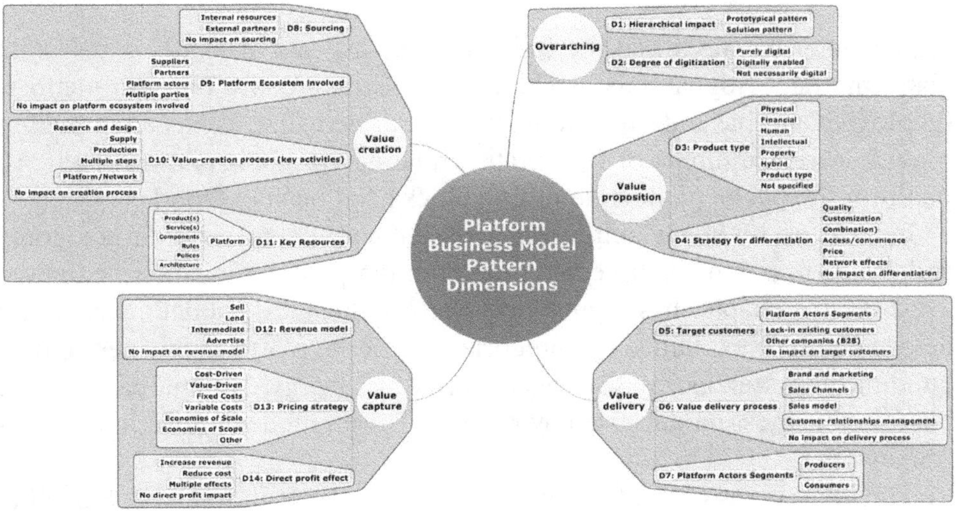

MSP BMP –> [industry/type] MSP BMP –> company MSP BM.

Different platform patterns, including MSP BMP, make the free offer possible. For instance, non-paying customers can be financed by another part of the BM or by another Customer Segment of the MSP BMP. In recent years with increasing of digitization of goods and services offered via Internet free digital product and service offers have exploded. They are usually used in Freemium Model and are related with fundamentally different economics of digital products and services, providing basic services free of charge and premium services for a fee (Teece, 2010; Osterwalder and Pigneur, 2010). In a free BM at least one substantial Customer Segment is able to continuously benefit from a free-of-charge offer. For example, creating and recording a song costs an artist time and money, but the cost of digitally replicating and distributing the work over internet is close to zero. Hence, an artist can promote and deliver music to a global audience over web, as long as he or she finds revenue streams, such as concerts and merchandising, to cover costs.

Long Tail BMP focuses on offering a large number of niche products, each of which being sold relatively infrequently (Osterwalder and Pigneur, 2010). Aggregate sales of niche items can be as lucrative as those of items within a traditional model, whereby a small number of bestsellers account for most revenues. Long Tail BM requires low inventory costs and strong MSPs to make niche content readily available to interested buyers.

CONCLUSION

This chapter has reviewed the following key aspects of platform research: platform strategy, dynamic capabilities and business models. The main platform typologies and basic definishions were described first. It has attempted to provide a brief summary of the literature relating to arguing platform strategy, platform life cycle, platform building blocks and business models.

It went on by suggesting a platform strategy categorization taxonomy and platform practical strategies of preventing platform disintermediation. Thus far, the chapter has argued that platform business model analysis and design could be explained by categorization of the platform business model building blocks taxonomy that was supported by strategy, dynamic capabilities, platform business models framework and business model innovation process framework.

The main types of platform business models were introduced. Selecting the platform business model and the platform strategy could reformulate more precisely what type of business the owner of the platform technology wants to become.

Thus far, the multi-sided platform business model pattern (MSP BMP) framework was introduced and discussed. We provide MSP BMP as a basic conceptual framework and knowledge management tool for describing, analyzing and interpreting non-price instruments used by digital platforms, especially platform intermediaries. Using the MSP BMP as the basis of the digital intermediaries' analysis and decision making, we update this pattern for each industry in next chapters. To conclude this section, platform business model design methodology was introduced. The chapter that follows moves on to consider the 14 dimensions of the BMP taxonomy for defining MSP BMP dimensions that could be used as a platform assessment tool.

REFERENCES

Afuah, A. (2004). *Business models: A strategic management approach*. New York: Irwin/McGraw-Hill.

Akaka, M. A., & Vargo, S. L. (2015). Extending the context of service: From encounters to ecosystems. *Journal of Services Marketing*, *29*(6/7), 453–462. doi:10.1108/JSM-03-2015-0126

Akaka, M. A., Vargo, S. L., & Lusch, R. F. (2013). The Complexity of Context: A Service Ecosystems Approach for International Marketing. *Journal of International Marketing, 21*(4), 1–20. doi:10.1509/jim.13.0032

Al-Debei, M. M., & Avison, D. (2010). Developing a unified framework of the business model concept. *European Journal of Information Systems, 19*(3), 359–376. doi:10.1057/ejis.2010.21

Alstyne, M., Parker, G., & Choudary, S. (2016, April). Pipelines, Platforms, and the New Rules of Strategy. *Harvard Business Review, 94*(4), 54–62.

Armstrong, M. (2006). Competition in two-sided markets. *The Rand Journal of Economics, 37*(3), 668–691. doi:10.1111/j.1756-2171.2006.tb00037.x

Blaschke, M., Haki, M., Riss, U., & Aier, S. (2017a). Design Principles for Business-Model-based Management Methods—A Service-dominant Logic Perspective. *Proceedings International Conference on Design Science Research in Information Systems and Technology (DESRIST).* 10.1007/978-3-319-59144-5_11

Blaschke, M., Cigaina, M., Riss, U., & Shoshan, I. (2017b). Designing Business Models for the Digital Economy. In G. Oswald & M. Kleinemeier (Eds.), *Shaping the Digital Enterprise. Trends and Use Cases in Digital Innovation and Transformation.* Springer International Publishing Switzerland. doi:10.1007/978-3-319-40967-2_6

Boley, H., & Chang, E. (2007). Digital ecosystems: principles and semantics. In *IEEE International Digital Ecosystems and Technologies Conference* (pp. 398-403). Cairns, Australia: IEEE.

Boudreau, K. (2010, October). Open Platform Strategies and Innovation: Granting Access vs. Devolving Control. *Management Science, 56*(10), 1849–1872. doi:10.1287/mnsc.1100.1215

Bouwman, H., De Reuver, M., Solaimani, S., Daas, D., Haaker, T., Janssen, W., . . . Walenkamp, B. (2012). Business Models Tooling and a Research. 25th Bled eConference Special Issue, Bled, Slovenia.

Brousseau, E., & Penard, T. (2006). The economics of digital business models: A framework for analyzing the economics of platforms. *Review of Network Economics, 6*(2).

Casadesus-Masanell, R., & Ricart, J. (2011). How to Design a Winning Business Model. *Harvard Business Review, 89*(1-2), 100–107. PMID:21510522

Coles, P., & Edelman, B. (2011). The Online Economy: Strategy and Entrepreneurship. Course Overview and Requirements. Harvard Buseness School.

Cusumano, M. A., & Gawer, A. (2002). The elements of platform leadership. *MIT Sloan Management Review*, *43*(3), 51–58.

Deimler, M., Lesser, R., Rhodes, D., & Sinha, J. (Eds.). (2013). *Own the Future: 50 Ways to Win from The Boston Consulting Group*. Wiley.

Edelman, B., Hu, P. (2016). *Disintermediation in Two-Sided Marketplaces.* Harvard Business School Technical Note 917-004, June 2016 (Revised March 2017).

Eisenmann, T. (2007). *Platform-Mediated Networks: Definitions and Core Concepts*. Harvard Business School, HBS 807-049.

Eisenmann, T., Parker, G., & Van Alstyne, M. (2006). Strategies for Two-Sided Markets. *Harvard Business Review*, 10.

Eisenmann, T. G., Parker, G., & Van Alstyne, M. (2011). Platform Envelopment. *Strategic Management Journal*, *32*(12), 1270–1285. doi:10.1002mj.935

El Sawy, O. A., & Pereira, F. (2013). *Business Modelling in the Dynamic Digital Space. An Ecosystem Approach*. Springer Briefs in Digital Spaces. doi:10.1007/978-3-642-31765-1

Ellison, N. B., Steinfield, C., & Lampe, C. (2007). The benefits of Facebook 'friends': Social capital and college students' use of online social network sites. *Journal of Computer-Mediated Communication*, *12*(4), 1143–1168. doi:10.1111/j.1083-6101.2007.00367.x

Evans, D. (2003). Antitrust economics of multi-sided platform markets. *Yale Journal on Regulation*, *20*, 325–381.

Evans, D. S., Hagiu, A., & Schmalensee, R. (2006). *Invisible engines: how software platforms drive innovation and transform industries*. Cambridge, MA: MIT Press.

Evans, D. S., & Mariscal, E.V. (2014). Market Definition Analysis in Latin America with Applications to Internet-Based Industries. *Journal of Law and Policy for the Information Society*, 9.

Evans, D. S., & Schmalensee, R. (2002). Some Economic Aspects of Antitrust Analysis in Dynamically Competitive Industries. *Innovation Policy and the Economy*, 2.

Evans, D.S., and Schmalensee, R. (2010). *Failure to launch: Critical mass in platform businesses*. Academic Press.

Evans, D. S., Schmalensee, R., Noel, M. D., Chang, H. H., & Garcia-Swartz, D. D. (2011). Platform economics: Essays on multi-sided businesses. Platform Economics: Essay on Multi-sided Businesses. *Competition Policy International*.

Evans, P. C., & Gawer, A. (2016). *The Rise of the Platform Enterprise A Global Survey*. New York: Center for Global Enterprise. Retrieved December 18, 2017, from: http://www.thecge.net/wp-content/uploads/2016/01/PDF-WEBPlatform-Survey_01_12.pdf

Frankenberger, K., Weiblen, T., Csik, M., & Gassmann, O. (2013). The 4I-framework of business model innovation: A structured view on process phases and challenges. *International Journal of Product Development*, *18*(3–4), 249–273. doi:10.1504/IJPD.2013.055012

Gassmann, O., Frankenberger, K., & Csik, M. (2014a). *The Business Model Navigator: 55 Models That Will Revolutionise Your Business*. Pearson Education Limited.

Gassmann, O., Frankenberger, K., & Csik, M. (2014b). Revolutionizing the business model. In O. Gassmann & F. Schweitzer (Eds.), *Management of the fuzzy front end of innovation*. Springer International Publishing. doi:10.1007/978-3-319-01056-4_7

Gawer, A. (2009). *Platforms, markets and innovation*. Edward Elgar Publishing. doi:10.4337/9781849803311

Gawer, A. (2011, Fall). What Managers Need to Know about Platforms. *European Business Review*.

Gawer, A., & Cusumano, M. (2002). *Platform Leadership: How Intel, Microsoft and Cisco Drive Industry Innovation*. Harvard Business School Press.

Gawer, A., & Cusumano, M. (2008). How companies become platform leaders. *Sloan Management*, *49*(2), 28–35.

Gawer, A., & Cusumano, M. A. (2014). Industry platforms and ecosystem innovation. *Journal of Product Innovation Management, 31*(3), 417–433. doi:10.1111/jpim.12105

Günzel, F., & Holm, A. B. (2013). One size does not fit all — Understanding the front-end and back-end of business model innovation. *International Journal of Innovation Management, 17*(01), 1–34. doi:10.1142/S1363919613400021

Hagiu, A. (2006). Pricing and commitment by two-sided platforms. *The Rand Journal of Economics, 37*(3), 720–737. doi:10.1111/j.1756-2171.2006.tb00039.x

Hagiu, A. (2014). Strategic Decisions for Multi-sided Platforms. *MIT Sloan Management Review, 55*(2).

Hagiu, A., Rothman, S. (2016, April). Network Effects Aren't Enough. *Harvard Business Review*.

Hagiu, A., & Wright, J. (2013, March). Do You Really Want to Be an eBay? *Harvard Business Review, 91*, 102–108. PMID:24730173

Hagiu, A., & Wright, J. (2015a). Marketplace or reseller? *Management Science, 61*(1), 184–203. doi:10.1287/mnsc.2014.2042

Hagiu, A., & Wright, J. (2015b, November). Multi-Sided Platforms. *International Journal of Industrial Organization, 43*, 162–174. doi:10.1016/j.ijindorg.2015.03.003

Hedman, J., & Kalling, T. (2003). The business model concept: Theoretical underpinnings and empirical illustrations. *European Journal of Information Systems, 12*(1), 49–59. doi:10.1057/palgrave.ejis.3000446

Hu, P. (2015). *Growing a marketplace? How to balance sellers and buyers using customer lifetime value*. Retrieved December 18, 2017, from: http://www.onlineeconomy.org/optimizing-two-sided-marketplace-growth-using-customer-lifetime-value/

Iansiti, M., & Levien, R. (2004). *The Keystone Advantage: What the New Dynamics of Business Ecosystems Mean for Strategy, Innovation, and Sustainability*. Boston: Harvard Business School Press.

Johnson, M. (2010). *Seizing the White Space: Business Model Innovation for Growth and Renewal*. Boston: Harvard Business Press.

Judah, S. (2016). *Disruption and the Digital Value of Business — the Rise of Digital Economics*. Gartner Report, ID: G00311450. Retrieved December 18, 2017, from: https://www.gartner.com/doc/3418917?cm_sp=swg-_-research-_-tail

Lecocq X., & Demil B. (2010). Business Models as a research program in strategic management: An appraisal based on Lakatos. *Management*, *13*(4).

Osterwalder, A. (2004). *The Business Model Ontology - A proposition in a design science approach* (Ph.D. thesis). University of Lausanne, Ecole des Hautes Etudes Commerciales HEC.

Osterwalder, A., & Pigneur, I. (2010). *Business Model Generation: A Handbook for Visionaries, Game Changers, and Challengers*. John Wiley & Sons.

Osterwalder, A., Pigneur, Y., & Tucci, C. L. (2005). Clarifying business models: Origins, present and future of the concept. Communications of the Association for Information Science, 16.

Parker, G., & Van Alstyne, M. (2005). Two-sided network effects: A theory of information product design. *Management Science*, *51*(10), 1494–1504. doi:10.1287/mnsc.1050.0400

Parker, G., & Van Alstyne, M. (2015). Platform strategy. In M. Augier & D. J. Teece (Eds.), The Palgrave Encyclopedia of Strategic Management. Available at http://www.palgraveconnect.com/esm/doifinder/10.1057/9781137294678.0794

Parker, G. G., Van Alstyne, M. W., & Jiang, X. (2017). Platform Ecosystems: How Developers Invert the Firm. *Management Information Systems Quarterly*, *41*(1), 255–266. doi:10.25300/MISQ/2017/41.1.13

Poli, R., Healy, M., & Kameas, A. (2010). *Theory and Applications of Ontology: Computer Applications Set: Theory and Applications of Ontology*. Springer.

Rappa, M. (2001). *Business models on the web: Managing the digital enterprise*. Retrieved December 18, 2017, from: http:www.digitalenterprise.org/models/models.html

Remane, G., Hanelt, A., Tesch, J., & Kolbe, L. (2017). The Business Model Pattern Database – A Tool for Systematic Business Model Innovation. International Journal of Innovation Management, 21(1).

Robertson, D., & Ulrich, K. (1998). Planning for Product Platforms. Sloan Management Review, 39(4).

Rochet, J.-C., & Tirole, J. (2003). Platform Competition in Two-Sided Markets. *Journal of the European Economic Association*, 1.

Rochet, J.-C., & Tirole J. (2004). Two-Sided Markets: Where We Stand. *Rand Journal of Economics*, *33*(4).

Rysman, M. (2009). The economics of two-sided markets. *The Journal of Economic Perspectives*, *23*(3), 125–143. doi:10.1257/jep.23.3.125

Shapiro, C., & Varian, H. (1999). *Information Rules: A Strategic Guide to the Network Economy*. Harvard Business Press.

Stabell, C. B., & Fjeldstad, O. D. (1998). Configuring value for competitive advantage: On chains, shops, and networks. *Strategic Management Journal*, *19*(5), 413–437. doi:

Sundararajan, A. (2016). *The sharing economy*. Cambridge, MA: MIT Press.

Taube, A. (2014). *The 20 Most Valuable Brands In The World*. Retrieved from: http://www.businessinsider.com/the-20-most-valuable-brands-in-the-world-2014- 5?op=1#ixzz3KMHUTYmY

Tauscher, K., & Laudien, S. (2017). Understanding platform business models: A mixed methods study of marketplaces. *European Management Journal*.

Teece, D. (2007). Explicating dynamic capabilities: The nature and microfoundations of (sustainable) enterprise performance. *Strategic Management Journal*, *28*(13), 1319–1350. doi:10.1002mj.640

Teece, D. (2010). Business Models, Business Strategy and Innovation. *Long Range Planning*, *43*(2), 172–194. doi:10.1016/j.lrp.2009.07.003

Teece, D. (2012). Next generation competition: New concepts for understanding how innovation shapes competition and policy in the digital economy. *The Journal of Law, Economics & Policy*, *9*, 97–118.

Teece, D. (2015b). Business Ecosystem, In M. Augier & D. J. Teece (Eds.), *The Palgrave Encyclopedia of Strategic Management*. Retrieved December 18, 2017, from: http://www.palgraveconnect.com/esm/doifinder/10.1057/9781137294678.0794

Teece, D. (2017a). *Dynamic Capabilities and (Digital) Platform Lifecycles*. Working Paper Series No. 17, Tusher Center on Intellectual Capital, Haas School of Business, University of California, Berkeley, USA.

Teece, D. (2017b). Business models and dynamic capabilities. *Long Range Planning*. (in press)

Teixeira, T., & Brown, M. (2016a). *Airbnb, Etsy, Uber: Acquiring the First Thousand Customers*. Harvard Business School Case 516-094, May 2016.

Teixeira, T., & Brown, M. (2016b). *Airbnb, Etsy, Uber: Growing from One Thousand to One Million Customers*. Harvard Business School Case 516-108, June 2016.

Thomson, J. (1967). *Organizations in action*. New York: McGraw-Hill.

Tiwana, A. (2014). *Platform Ecosystems: Aligning Architecture, Governance, and Strategy*. Morgan Kaufmann Publishers.

Van Alstyne, M. W., Parker, G. G., & Choudary, S. P. (2016). 6 Reasons Platforms Fail. *Harvard Business Review*. Retrieved December 18, 2017, from https://hbr.org/2016/03/6-reasons-platforms-fail

Vargo, S. L., & Lusch, R. F. (2016). Institutions and axioms: An extension and update of service dominant logic. *Journal of the Academy of Marketing Science*, *44*(1), 5–23. doi:10.100711747-015-0456-3

Yablonsky, S. (2013). Multisided Platforms and Markets: Basic Approaches, Concepts and Practices [in Russian]. *The Russian Management Journal*, *11*(4), 57–78.

Yablonsky, S. (2014). E-finance innovation services in Russia. *International Journal of Entrepreneurship and Innovation Management*, *8*(5), 523–551.

Yablonsky, S. (2016). Intermediaries in E-Commerce. In *Encyclopedia of E-Commerce Development, Implementation, and Management* (pp. 48–73). IGI Global. doi:10.4018/978-1-4666-9787-4.ch005

Zott, C., & Amit, R. (2015). Business Model Innovation: Toward a Process Perspective. In The Oxford Handbook of Creativity, Innovation, and Entrepreneurship. OUP.

Zott, C., Amit, R., & Massa, L. (2011). The Business Model: Recent Developments and Future Research. *Journal of Management*, *37*(4), 1019–1042. doi:10.1177/0149206311406265

Chapter 2
Digital Platforms:
From Technology to Leadership Platforms

ABSTRACT

This chapter focusses on the necessary platform components like technology, leadership, talent and skills, delivery, trust, marketing, ecosystem, and business models required to support the capabilities of digital business innovations. The multidimensional platform organization and management is discussed.

INTRODUCTION: DIGITAL BUSINESS PLATFORMS

'Digitalization' can be defined as "the use of digital technologies to change a business model and provide new revenue and value-producing opportunities; it is the process of moving to a digital business." (Gartner, 2017).

Today the adoption of digital innovation technologies redefines e-Commerce and e-Business. E- commerce – the interaction (buying and selling) among businesses and people conducted electronically on the Internet – is transformed in digital commerce – the interaction (buying and selling) among businesses, people and things for products and/or services via digital channels that results in a transaction of value to the customer. Customer value is based on a combination of factors such as a great experience, low price, timeliness, ease of use, policies, etc.

DOI: 10.4018/978-1-5225-5457-8.ch002

Jorge Lopez (2014) explains the transition from e-business to digital business as 'presence and integration of things, connected and intelligent, with people and business'. An interconnection of such kind requires a network that brings everything together and a cloud infrastructure providing all services necessary for coping with every demand of today's business agility (Blaschke et al., 2017a.b).

Digital business (see Figure 1) is a creation of new business designs by blurring the boundaries between the digital world and the physical one due to the convergence of people, businesses and things. It promises an unprecedented convergence of a) people, b) businesses and c) things that changes existing platform business models and creates new revenue opportunities and platform designs (Lopez, 2014, 2017; Cearley et. al, 2017; Yablonsky, 2016, 2017). In digital business, business and technology are inextricable. To model a business, managers must understand what is needed for a business to achieve its goals, and how these goals translate into actionable strategy. For example, turning products into digitalized services that generate a steady revenue stream implies significant changes within all the other components.

A digital business requires much more than technology (for example, leadership, talent, skills and new business models). To architect a digital business platform, organizations must lead their business to take a business-driven, outside-in approach. Digital business platforms empower flexible and dynamic digital business. Unlike a traditional business, with a clear inside and outside, a platform provides business with a foundation where resources can come together in different combinations - to create value. Some resources may be inside, permanently owned by the company; some may be shared; and some can come from an outside ecosystem. The value comes largely from dynamic connection of those resources and the actors, and the network effects between them.

Platform design facilitates matches among providers and consumers ("users") or, in other words, a creation or exchange of goods, services and social currency, so that all participants can capture value.

Platforms offer unique opportunities to engage members of a business ecosystem to exchange value (Blosch and Burton, 2016). However, they require an entirely new approach to strategy, management and architecture. The answers to the question are not limited to digital business platform technology decisions, but also include businesses, markets, economics, partners, customers, human capital, processes, information and technologies.

Figure 1. Digital business actors

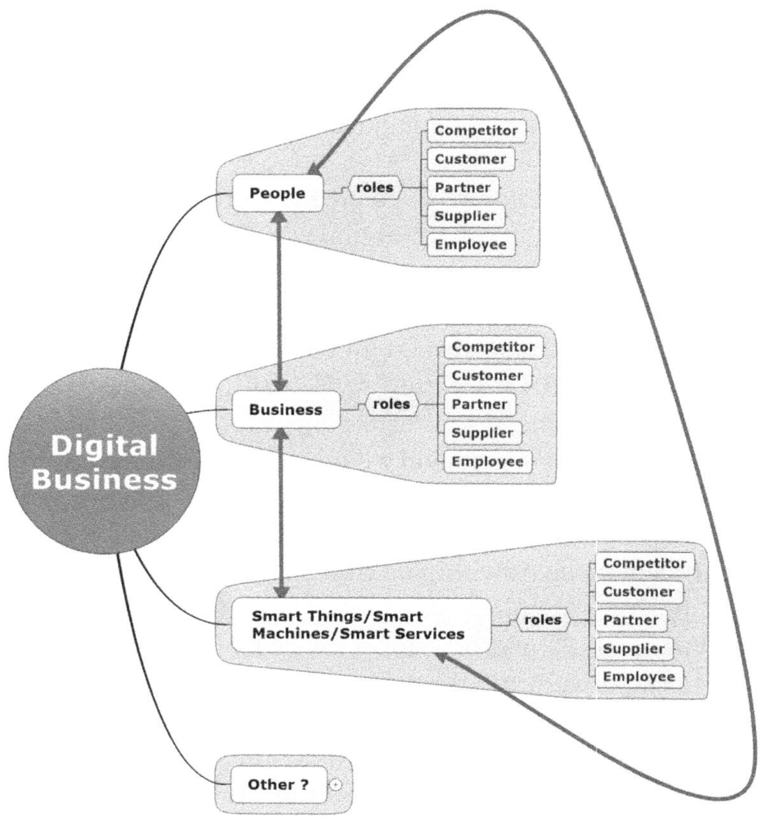

Digital platform disruption is a process of changing multi-sided markets through digital capabilities, channels and assets. Digital business innovation forms disruptive platform effects. To manage digital platform business models and multi-layered platform business ecosystems, companies are creating a digital business platform stack to share critical assets. The variance in company's digital business performance is a function of differences in the platform resources and capabilities of competing companies.

Definition 1 (Burton and Basiliere, 2016).

A *digital platform* is a business-driven framework that allows a community of partners, providers and customers to share and enhance digital processes and capabilities or to extend them for mutual benefit. This framework allows

Figure 2. Industry vision of digital transformation into a scalable digital platform business

different combinations of business models, leadership, talent, delivery and IT infrastructure platforms that power digital business ecosystems.

Platform strategy becomes extremely complex as firms consider dynamic interactions of a multi- layered business ecosystem (Teece, 2012).

The concept of '*Platform innovation*' can be widely defined as architecting new platform strategies and business models, making and promoting new platform products and services, developing new platform processes to facilitate platform activities, to interact with platform actors and to design new platform structures for industry institutions.

The concept of *'Industry platform innovation'* entails changes to infrastructure, standards and rules that enable transactions between multiple firms (Leiblein, 2015). Industry platform innovation vision shows what platform business could look like if enterprises used technology innovations to their full potential.

The industry vision framework gives companies a structure to refine and fill out their thinking about digital business. When their ideas have matured, they can use the framework to explain the concept to the rest of their organizations and to their partners. Leaning on Moyer (2016), an industry platform innovation vision framework consists of four parts: concept, capabilities, assets and research (see Figure 2).

Basic Platform Innovation Dimensions

We have selected the following key dimensions of platform innovation.

1. *Three basic platform dimensions*:
    a. Components,
    b. Rules,
    c. Strategies.
2. *Four additional platform players "actors/sides" dimensions, proposed as follows*:
    a. Producers,
    b. Consumers,
    c. Providers,
    d. Owners.
3. *Innovation technology dimensions* (innovative dynamic digital key elements). The term has become popular to describe the convergence of innovation technologies. In particular, we refer to different approachers with different stage of granularity: from simplified SMACIT (social, mobile, analytics, cloud, and the Internet of Things) and SAP Digital Key Elements from BMDI: SAP Business Model Innovation Methodology (Oswald and Kleinemeier, 2017) through IDC Innovation Accelerators and third-platform concept (IDC, 2017) to dynamic Gartner Top 10 Strategic Technology Trends (see Figure 3) for every year (Cearley et al., 2017). For example, we can use the following set:
    a. IoT.
    b. Cloud computing.
    c. Big data.
    d. Artificial Intelligence (AI).
    e. Mobile.
    f. Others.

Figure 3. Gartner Top 10 Strategic Technology Trends for 2017

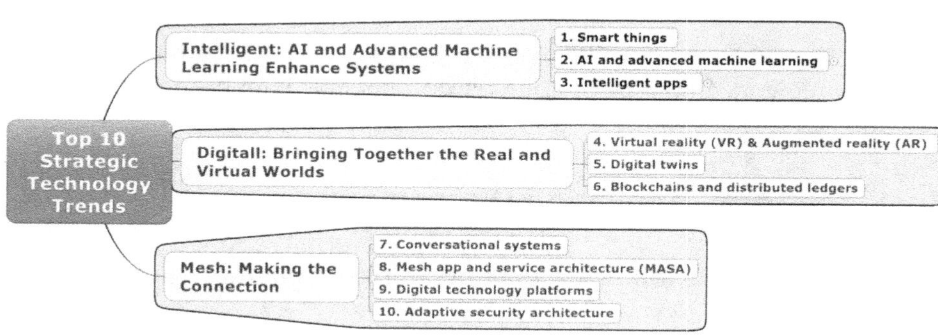

The general framework of platform innovations is shown on Figure 4.

Figure 4. Platform Innovations and Capabilities

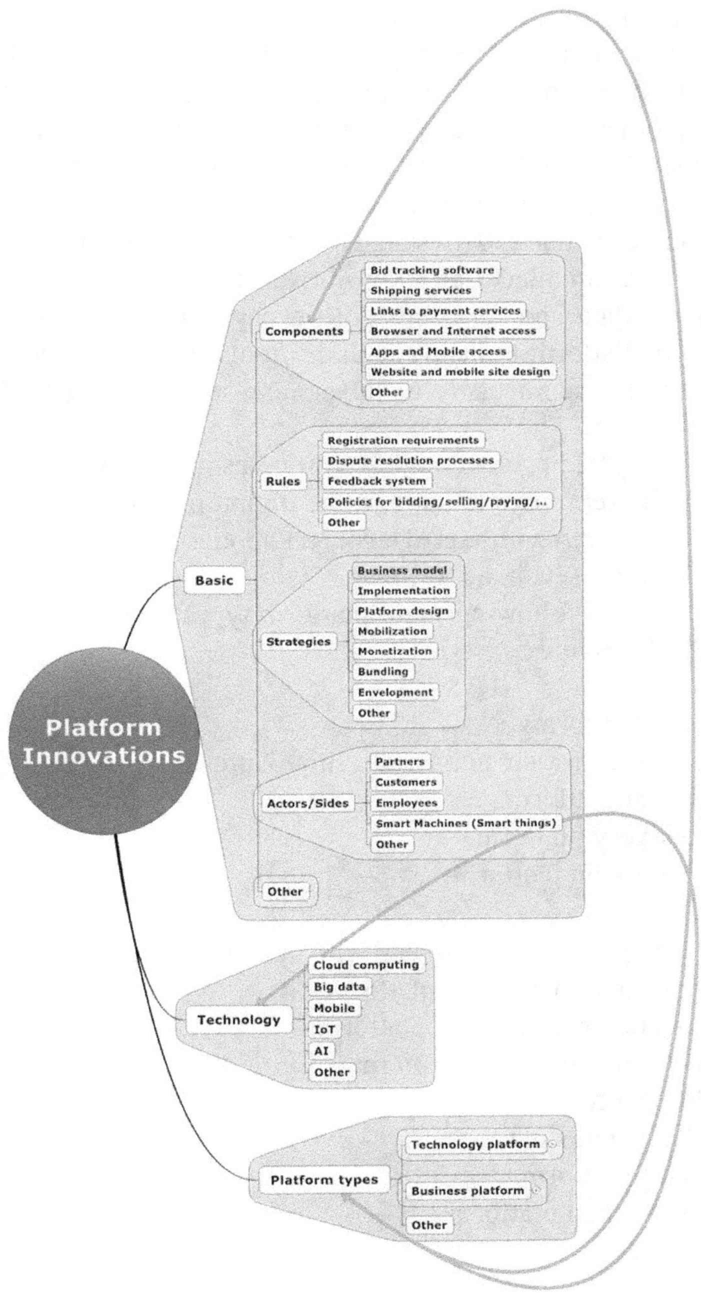

DIGITAL PLATFORM STACK

Every firm has tangible and intangible business assets (processes, human talent, technologies, infrastructure and information) that can be used or developed to bring the digital business platform to life (Burton & Basiliere, 2016). As it was mentioned in Chapter 1 multisided platform (MSP) is usually a subset of components (hardware, software, services), rules (technical standards, protocols for information exchange, policies, and contracts that govern transactions) and strategy/business models building blocks employed by users in most of their transactions.

Hunter and Coleman (2016) stated that "in the context of digital business, a platform is an architecturally innovative means of *sharing assets* such as algorithms, data and functions with ecosystems of people, businesses and things. A platform typically includes tools, functions and support for multiple customer segments (e.g., buyers and sellers; students, faculty and administrators)".

With new digital business models, resources may be added or combined in new and different ways to support the digital platform strategy — or an organization may start with a set of resources applied across the entire business or apply specific assets in a few areas.

We propose the following main platform types related with the Digital Platform Stack (Figure 5):

1. Business Platforms
    a. Business Model and Leadership platform.
    b. Talent platform.
    c. Delivery platform.
    d. Promotion platform.
    e. Other.
2. Technology platforms – related with innovation technology dimensions
    a. Information systems platform.
    b. Customer experience platform.
    c. Data and analytics platform.
    d. IoT platform.
    e. Ecosystems platform.
    f. Trust platform.
    g. Integration platform.
    h. Other.

Business Platforms

Business Model and Leadership Platform

With platform business models, platform assets may be added or combined in new and different ways to support the digital platform strategy. Delivering digital platform requires new capabilities to enable, support and evolve digital business (Burton & Basiliere, 2016). Platform Business Model (BM) *management* is an important managerial function (Osterwalder, Pigneur, 2010; Hagiu, 2014). It involves the ongoing monitoring of the activities that encompass the BM and of the incentives of all the stakeholders who participate in the BM. BM management thus considered could be viewed as part of a firm's ordinary capabilities (in terms of the day-to- day performance

Figure 5. Digital Platform Stack (adopted from Le Hong et al., 2016)

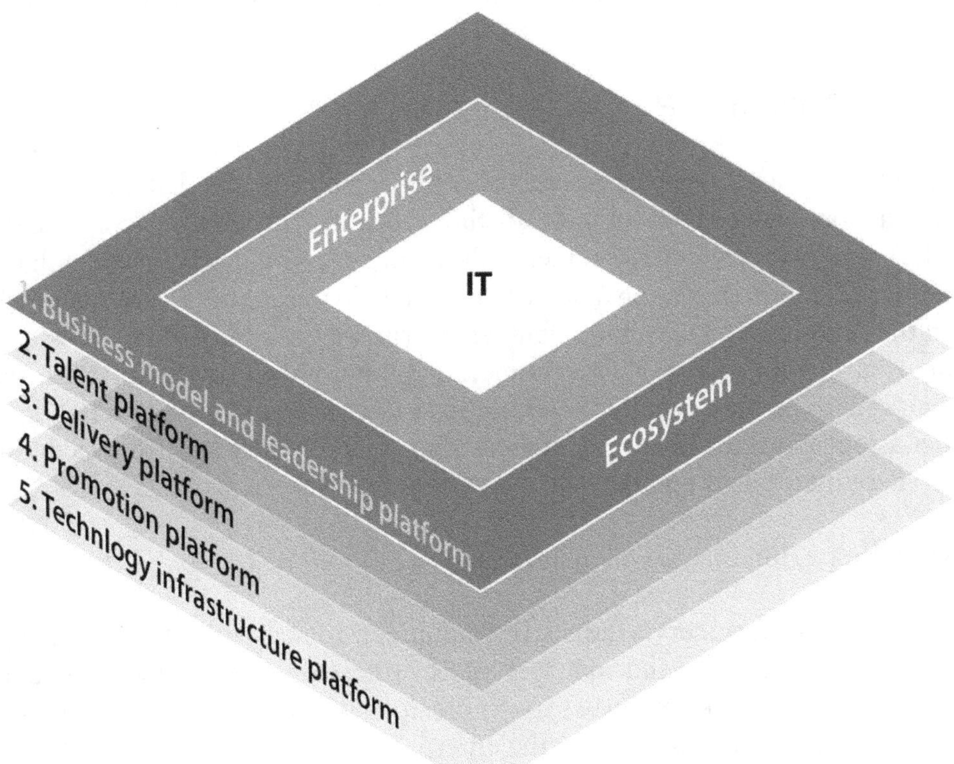

of activities), but also requires dynamic adaptation and transformation, and thus links to the dynamic capabilities framework.

Business Model and Leadership, as well as Talent platform, are related more with platform capabilities. Their goal is to facilitate knowledge exchange in Business Model and Leadership environments and to offer affiliated actors the opportunity to access large intra- or ecosystem communities of actors with experiential, educational, or professional knowledge in company's diverse geographical and disciplinary fields (Boudreau, 2010; Boudreau et al., 2011; Colombo et al. 2013; Colombo et al., 2015; Evans and Gawer, 2016). The key roles of a Business Model and a Leadership platformare described as to collect dispersed sources of knowledge, to recombine the collected knowledge in order to empower innovation and management, and to transfer it to new technological and organization contexts.

Delivery Platform (DP)

A delivery business platform creates value primarily by enabling direct interactions between two (or more) distinct types of affiliated delivery consumers.

Talent Platform (TP)

A talent business platform creates value primarily by enabling direct interactions between two (or more) distinct types of affiliated talent consumers. The talent platform is at the center of the enterprise's relationship with talent. How can enterprise acquire and keep top talent in the age of digital business? In the relationships with top talent, enterprise can use marketers' tools and analyses. Seeing talent as a customer, and employment-by-talent platform as a brand promise fulfilled, can improve talent acquisition and retention. Persistent shortfalls in key talent areas show that enterprises have to act now to adapt their talent approaches to the digital world.

Definition (Hunter and Coleman, 2016)

A talent platform serves the need of multiple customer segments — including enterprise executives and managers, HR professionals and recruiters, and potential or current employees — to create and maintain engagement and evolving relationships between the enterprise and its contributors, internal and external. A key function of the talent platform is data capture and analysis related to talented individuals and talent pools — before, during and after their employment by the enterprise. The talent platform specifically supports an approach to talent that recognizes and leverages the "customer decision life cycle" of explore, evaluate and engage (see figure below). Treating talent as a customer whose relationship with the enterprise includes a mix of exploration, evaluation and engagement over time is more realistic and fruitful for all involved than treating the acquisition of talent as a transaction.

Promotion Platform (PP)

A promotion business platform creates value primarily by enabling direct promotion interactions between two (or more) distinct types of affiliated platform participants – consumers, producers, and providers. It provides creating internally managed outbound messages and externally inbounded messages by platform participants themselves. A promotion platform influences platform participants – consumers, producers, and providers – to submit multimedia messages, provide different types of activities and promote them on social media and platform ecosystems.

Technology Platforms

To manage the new digital business models and business ecosystems, companies have to create a digital business platform to share critical assets. Coltman and Queiroz (2015) stated that a firm's strategy at the corporate level refers to the set of choices about how to compete across the different businesses that constitute the corporate profile. At this level, strategic decisions are made about how technological platforms can support levels of data sharing and business process standardization across business units.

Corporate level strategy is reflected in the corporate technological platforms that include data, hardware, network, applications and management services that are shared by business units. Thus, a core function of technological platforms is to provide the foundation that allows BUs business units to leverage common factors of production and to promote process synergies where the joint value creation is greater than the sum of the value created by individual businesses. Realizing these synergies requires continuous efforts of alignment to ensure integration of corporate strategy and corporate technological platforms capabilities.

Figure 6. Digital Platform Stack with Capabilities

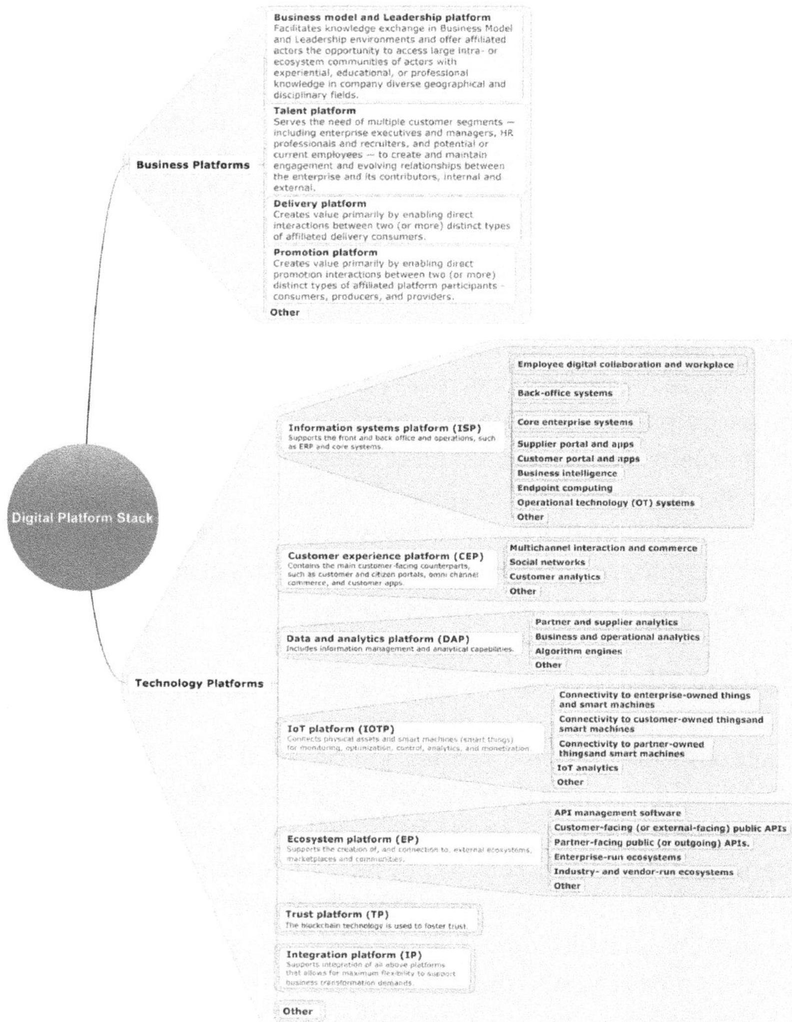

Digital technology platforms are described through the lens of applications and business capability components they support (see Figure 6). There are multiple layers to these technological platforms which are necessary to effectively implement and integrate it in practice (LeHong et al., 2016). Business platforms could be supported by technology platforms in seven overlapping areas (see Figure 7):

Figure 7. Technological platforms overlapping (adopted from Le Hong et al., 2016)

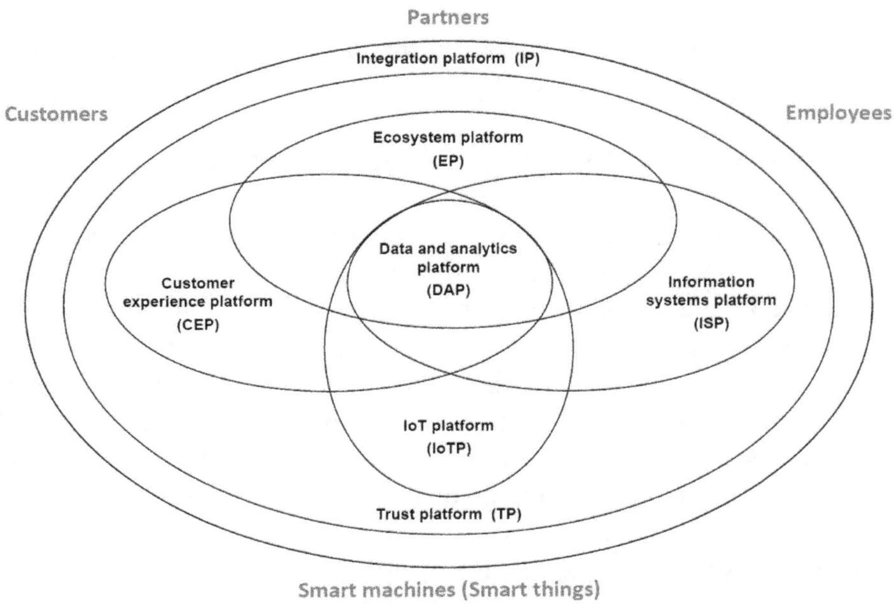

1. **Information Systems Platform (ISP):** Supports the front and back office and operations, such as ERP and other core systems.
2. **Customer Experience Platform (CEP):** Contains the main customer-facing counterparts, such as customer and citizen portals, omni channel commerce, and customer apps.
3. **Data and Analytics Platform (DAP):** Includes information management and analytical capabilities. Data management programs and analytical applications fuel data-driven decision making, and algorithms automate discovery and action.
4. **IoT Platform (IOTP):** Connects physical assets and smart machines (smart things) for monitoring, optimization, control, analytics, and monetization. Capabilities include connectivity, analytics and integration to core and OT systems.
5. **Ecosystems Platform (EP):** Supports the creation of, and connection to, external ecosystems, marketplaces and communities. API management, control and security are its main elements.
6. **Trust Platform (TP):** A blockchain technology used to foster trust.
7. **Integration Platform (IP):** Supports the integration of all above platforms that allows the maximum flexibility to support business transformation demands.

Information systems platform (ISP)

The purpose of this platform is to run the back-office and core enterprise systems. The information systems platform contains:

- **Employee Digital Collaboration and Workplace:** Email, telephony and video, file storage and sharing, collaboration tools, productivity tools, employee app stores and personal employee apps are all part of the tools used by employees in the digital workplace.
- **Back-Office Systems:** These include finance, HR, purchasing, billing and other vital back-office systems. Systems of record for master data are also included. Usually, these back-office capabilities are delivered by an ERP system and "bolt-ons". For certain domain areas, like HR, these systems are shifting to the public cloud to support the pressing need to better attract and retain talent or optimize procurement or logistics.
- **Core Enterprise Systems:** Each industry has a set of core systems (e.g., core banking and core merchandising). The terms "core vertical systems", "industry applications" and "vertical IT systems" are also used to describe core systems in this report. Examples include provisioning systems in telecom, actuarial systems in insurance, and supply chain and warehouse systems in industrial sectors. Industry-specific versions of ERPs and best-of-breed industry solutions often provide these capabilities.
- **Supplier Portal and Apps:** These are used by vendors to transact, view and change orders electronically. Mobile apps and websites usually deliver the portal user experience, while an underlying layer uses industry standards to conduct transactions and connect to supply chain hubs. Supplier portals can evolve to become full-scale ecosystems.
- **Customer Portal and Apps:** Customers use them to gain information, get customer service, apply for services, place orders and transact, and/or view status. For public- sector enterprises these are often citizen portals and apps, and for other enterprises it can be a website or app. For retailers it is often their e-commerce site and apps. Customer portals can evolve to become full-scale ecosystems.
- **Business Intelligence:** This includes reporting, dashboards and online analytical processing (OLAP) analysis of back-office and core system data. These tools are used for historical reporting to enable advanced

decision making. The presentation layer can be anything from web and mobile to sophisticated interactive visualization. This data is becoming available to even wider corporate audiences (self-service portals), allowing more sophisticated analysis, such as forecasting. They can be deployed as a part of other systems (e.g., HR, CRM, ERP) or across various applications.

- **Endpoint Computing:** The endpoint devices and the security and management tools that support them are part of this capability. The placement of endpoint computing in Figure 2 denotes the progressing convergence between endpoint computing and IoT.
- **Operational Technology (OT) Systems:** Although not part of the traditional IT landscape, these operational systems directly control and monitor equipment and assets. For example, in manufacturing, they can be industrial control systems or supervisory control and data acquisition (SCADA); in utilities, they can be energy management, substation automation or SCADA systems; and in healthcare they are the clinical systems and technologies. Manufacturing execution systems, fleet telematics and telecom network equipment are also examples of operational technologies.

Customer Experience Platform (CEP)

The purpose of this platform is to run the technology that enables interaction and commerce with customers (B2C, B2B, B2B2C) and citizens.

The customer experience platform contains:

- **Multichannel Interaction and Commerce :** Preferably, these customer activities are API-based and can be combined to form any customer-facing process across any channel. Multichannel interaction and commerce can be B2C, B2B, B2B2C and citizen-facing.
- **Social Networks:** Enterprises ought to be able to participate in social networks for community interaction and transactional (e.g., commerce and payment) functions. The customer experience platform needs to be able to support this aim by enabling the posting of multimedia data (feeds, graphics, photos, videos etc.) and by receiving data (e.g., customer questions from Facebook, reviews and opinions etc.). It can also provide a mechanism for employees to engage with customers or participants to improve the experience.

- **Customer Analytics:** Data, metadata and aggregated data from social networks form customer profiles and is used for customer analytics accessed from all internal and external sources (e.g., CRM, purchased data, social networks, ERP, mobile app, etc.). Enterprises will use analytics ranging from simple historical analysis and segmentation to on-demand diagnostic, predictive and prescriptive analytics and advanced social sentiment. The output/results of customer analysis can be transferred to any other parts of the technology platform.

Usually a customer experience platform is connected with *Customer portal and apps* and *Back-office systems* from the *Information Systems Platform.*

Customer experience platform is the gateway through which other platforms (e.g., IoT and information systems) provide communication and value to the customer. Inbound data, insight, requests and commerce, as well as outbound data and fulfillment of services go through the customer experience platform. Through this the enterprise provides "a single view of the customer" as well as the customer is supported with "a single view of their enterprise". For example, customers should be able to walk into a branch, store or service center and ask about transactions and service that were done on a mobile app. This requires fluid integration of products/services/channels, and an outside-in, customer-centric mindset for development.

Data and Analytics Platform (DAP)

Data infuses all platforms. Data that supports partner and supplier analytics can originate from internal sources as well as partner/supplier/third-party sources. This in addition includes data obtained from the expanding landscape of data brokers and open datasets. Therefore, data and analytics platform holds a central position at the intersection of all other platforms. The main purpose of this platform is to provide the following items:

- Context for compliance and reporting.
- Real-time event analysis and adjustments to enterprise processes.
- Data and models that allow decisions to be made.
- Algorithms that automate decisions and prescribe courses of action that can be executed in all of the platforms.

Each part of a data and analytics platform can deliver insight that is descriptive, diagnostic, predictive and/or prescriptive.

Data, analytics and algorithms are essential to the digital business platform, and should be integrated by integration platform services to permit other platforms to use external and internal data and analytics to execute its functions. To democratize data processing and visualization this platform should include self-service for increasingly wider enterprise constituencies.

Data quality is a key success factor of the adoption of this type of platform. Data and analytics strategy and governance framework is a necessary component of data and analytics platforms.

A data and analytics platform includes:

- **Partner and Supplier Analytics:** That should be able to analyze the services rendered and products delivered by vendors and partners, to insight supplier fill rates, broker- dealer/agent performance, partner fraud analysis and some other tools that will help the enterprise, as well as to provide multitier supply chain visibility and analytics.
- **Business and Operational Analytics:** resulting from operations and business functions is critical for all parts of the enterprise. Sales, production and process analytics are the most common types of analysis. Most of the data usually comes from the core system and back-office sources, which may be supplemented by other sources.
- **Algorithm Engines:** Algorithm and analytics modelling workbenches create and then run the algorithms used in all parts of the enterprise (e.g., customer offer engines, fraud detection engines and predictive maintenance). Although centrally positioned in the figures, algorithms are used throughout all of the platforms.

A data and analytics platform should be connected with *Business intelligence* from the *Information Systems Platform.*

IoT Platform (IoTP)

Definition

The Internet of Things is defined as:

A network of dedicated physical objects (things) that contain embedded technology to sense or interact with their internal state or the external environment. This excludes general-purpose devices such as smartphones, tablets and PCs.

The purpose of this platform is to connect endpoints such as physical assets, consumer [smart] things, and in the nearest future smart machines.

The platform enables monitoring, management, services and security of these things. The platform also brings together the things connected to it with existing IT and OT systems to enable further value.

An IoT platform contains

- **Connectivity to Enterprise-Owned Things and Smart Machines:** Any enterprise has a whole set of heterogeneous assets that it owns and uses such as manufacturing equipment, medical equipment, vending machines and store equipment, etc. The IoT platform provides connectivity to this equipment, and should be able to handle many different data and wireless standards.
- **Connectivity to Customer-Owned Things and Smart Machines:** Enterprises should be able to connect to their customer's things. At the same time consumer things and smart machines (e.g., appliances and electronics), cars and many other consumer-owned things should be able to connect to an enterprise's IoT platform.
- **Connectivity to Partner-Owned Things and Smart Machines:** Any enterprise uses assets that it leases or that are operated by partners. An IoT platform should also perform as a gateway to data from partner-owned things and smart machines.
- **IoT Analytics**: An IoT platform can be used in two ways to support analytics. First, it can act as a gateway to analytics and storage capabilities that exist in the data and analytics platform. Second, the IoT platform can to do some or all of the data monitoring, storage, analytics and algorithms in the IoT platform. In most cases IoT analytics will be used as a mix of these two.

An IoT platform should be connected with *OT systems, Core systems, and Endpoint computing* from the *Information Systems Platform.*

Ecosystem Platform (EP)

The purpose of this platform is to empower an enterprise to create value from the outside business ecosystems - a whole business network (company, customers, partners, and global competition). This requires the ability to make assets like data, algorithms, transactions and business processes available to external business ecosystems through APIs; to construct ecosystems that an enterprise can host to connect new partners and developers, and an ability to connect to industry ecosystems such as marketplaces, supply chain hubs and financial networks. It also requires organizations to consider the relationship between internal, private and public APIs to unlock new revenue opportunities from existing services and information. We provide a generalization of modern structures of ecosystems (Blosch, Burton, 2016; Oswald and Kleinemeier, 2017) creating a platform ecosystem structure (see Figure 8).

Ecosystem platform contains:

- *API management software* to provide proper management, security and governance for an enterprise's sets of ecosystem's APIs in order to interconnect employees, customers and partners. Banks, retailers, governments and any kind of enterprise can expand their capabilities into the digital world via APIs. An API management software can make APIs available to business ecosystems and integrate with APIs outside of the enterprise. It serves both incoming (from external providers) and outgoing (from the enterprise's public-facing APIs) APIs, and it provides a mechanism to inspire development of new services and to manage usage of APIs.
- *Customer-facing (or external-facing) public APIs and partner-facing public (or outgoing) APIs.* These APIs are not applications or apps — they are key functionalities to be used in external applications, apps and websites.
- *Enterprise-run ecosystems* could be built by enterprises as commercial or community- based platform ecosystems. For example, an enterprise may want to create a buyer/seller/third-party platform marketplace for products/services (e.g., eBay, Amazon, Uber or Airbnb). The ability to create social graphs, feeds, monetization engines and algorithms that link supply with demand (e.g., recommendation engines) are some of the main capabilities of such platforms.
- *Industry- and vendor-run ecosystems* are provided in some industries as different hubs, communities and networks that bring together players in the value chain. For example, retail marketplaces, B2B marketplaces, credit card networks, payment networks and coupon clearinghouse networks, and supply chain hubs are external ecosystems run by an industry consortium or third party, technology vendors or suppliers and customers. The technologies that connect and allow interaction and commerce to occur are part of the capabilities of an ecosystem platform.

Figure 8. A platform ecosystem structure with capabilities

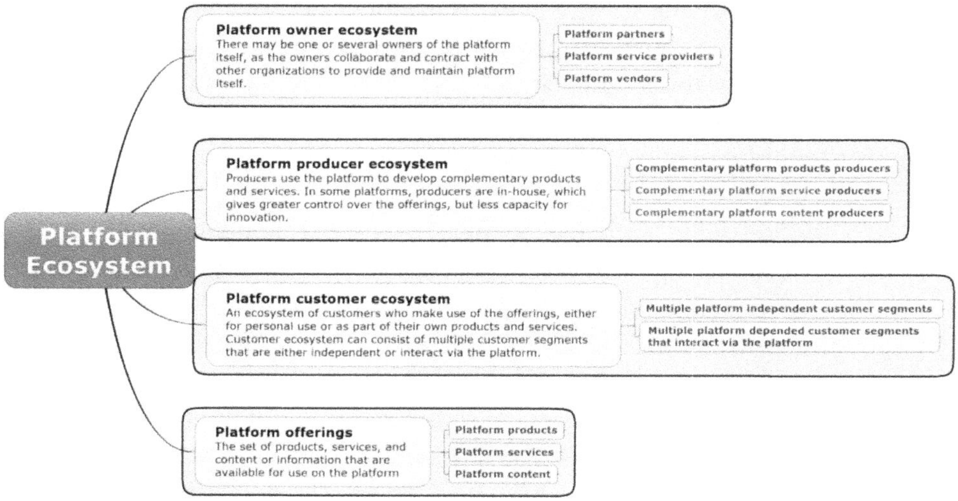

Customer portal and apps plus s*upplier portal and apps* are one of an enterprise's first outreach to external ecosystems.

An ecosystems platform should be linked to all other platforms directly or through integration platform. All other platforms will use it as a way to digitally interact and/or transact with the external world. In many ways, the ecosystems platform serves as a gateway and host to the external world. Many enterprises do not decide to build their own ecosystem. However, mostly all enterprises would like to have access to external APIs or to expose their own one publicly.

Trust Platform (TP)

A blockchain technology that makes it possible to create a digital ledger of transactions and share it among a distributed network of computers is used to foster trust in the currency and transaction mechanism not backed by any government or traditional institution.

Integration Platform (IP)

An Integration platform integrates different stack platforms by information and semantics integration. Ideally it should provide platform's interoperability, to work with other platforms and ecosystems, present or future, in either implementation or access, without any restrictions. The goal is to create an interoperable set of services that can be brought together to create applications, apps and workflows. This creates a synergetic collection of technology capabilities and components that form a platform. We could think of services as Lego blocks that can be easily rearranged to meet any need. The openness and composite nature of a platform stack is ideally suited to the external-facing capabilities required by new digital business processes, moments and models.

IP should provide the following types of interoperability:

- *Syntactic interoperability* to permit information exchange between platforms using specified data formats and communication protocols.
- *Semantic interoperability* to allow automatic meaningful and accurate interpretation of the exchanged information in order to produce useful results as defined by the end smart machine users of both sides.
- *Cross-domain interoperability* for multiple social, organizational, political, legal entities working together for a common interest and/or information exchange in business platforms stack.

Different users, such as integration specialists, line-of-business (LOB) developers, SaaS administrators and other IT skilled personnel ("ad hoc" integrators), as well as business users (citizen integrators) should have access to the integration platform capabilities appropriate for their profiles (Guttridge et al., 2017).

Successfully building a digital business platform strategy will hinge on an enterprise's ability to define an approach to integration that allows maximum flexibility to support shifting business demands.

A market for "out of the box" IP offerings is developing, but the provider landscape is extremely fragmented and the maturity of the rapidly evolving technologies may not be consistent across offerings' components (Guttridge et al., 2017).

There are multiple layers for technology platforms´ description which are necessary to effectively implement and integrate it in practice. There are such additional areas of description (Burton and Basiliere, 2016) as follows:

- *An infrastructure and operations perspective (e.g., data centers, networking and cloud).* All of the capabilities in the platform can be sourced from any combination of internal resources or external partners (insourcing, outsourcing, as-a-service and cloud sourcing).

- *A data management and retention perspective* that is covered by policy.
- *A security, risk and compliance perspective,* needed for building a business technology platform to understand risk tolerance and effectively implement security controls to balance exposing information to exploit digital business moments while mitigating the risks that can be created. In many industries there are regulatory compliance requirements that may influence the platforms (e.g., the need to keep customer data exclusively inside the enterprise as opposed to keeping it with ecosystem partners).

All capabilities in the technology platforms´ stack can be sourced from any combination of internal resources or external partners (insourcing, outsourcing, as-a-service and cloud sourcing).

GENERAL FRAMEWORK FOR SYSTEMATIC PLATFORM INNOVATION

The digital business platform stack is described through the lens of technology and business platform layers, applications and business capability components. It's intended to provide a higher level overview of the key capabilities necessary to assemble a digital business platform stack. Platforms can very quickly become highly complex. So enterprise platform architecture must be closely involved in the planning and evolution of the platform, along with the design of the technical architecture, delivery model and strategy. A successful platform strategy involves creating an offering that is valuable to the ecosystem, innovative and dynamic, while evolved in a measured, coordinated way. Care must be taken to balance innovation and execution (Blosch and Burton, 2016).

Methodology

Drawing on existing knowledge in the field of business model patterns (Osterwalder and Pigneur, 2010; Osterwalder et al., 2014; Blosch and Burton, 2016; Blaschke et al., 2017 a,b) and platform research (Evans, and Gawer, 2016) we argue that a meta-structuring perspective serving through the platform landscape represents an important contribution that remains missing. The objective of this research was to create such a meta-structuring perspective.

Figure 9. Platform business model pattern dimensions

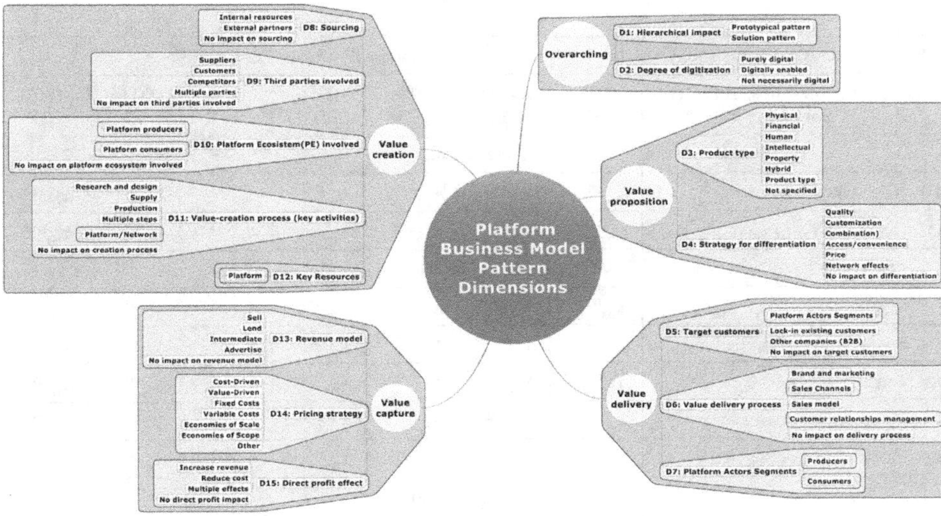

Therefore, we apply an ontology/taxonomy building methodology to create a consistent structure of the platform innovation ontology.

First we have to define meta characteristics of the platform innovation dimensions. A platform business model is a useful lens for understanding the company's platform underlying logic because it explains what value is provided, how this value is created and delivered, and ow profits can be generated therefrom. This concept helps to understand the capturing value from technological innovations and platforms (Chesbrough and Rosenbloom, 2002; Osterwalder and Pigneur, 2010; Osterwalder et al., 2014), the boundaries of a firm (Zott and Amit, 2010), and create a direct connection between business strategy and business processes (Al-Debei and Avison, 2010). Therefore, using *platform business model pattern dimensions framework* (Chapter 1) we replace *basic platform dimensions* with *expanded platform business model pattern dimensions* (see Figure 9).

Main technology dimensions and platform type dimensions could be enlarged with *business/technology platform dimensions*. As a synthesis of different views that technology analysts, researchers, and practitioners are commonly using today, we analyze and design platform innovation dimensions through a minimalistic, object-oriented, and functional representation based on seven key technology platforms (see Table 1).

The level of granularity depends on the needed level of detalization (see Table 2).

Table 1. Matrix form relationship between business model and technology platforms dimensions

Business Platform Dimensions	Technology Platforms Dimensions						
Platform Business Model Pattern Dimensions	**DT1: ISP**	**DT2: CEP**	**DT3: DAP**	**DT4: IoTP**	**DT5: EP**	**DT6: TP**	**DT7: IIP**
Overarching							
D1: Hierarchical impact	A_{D1}^{DT1}	A_{D1}^{DT2}	A_{D1}^{DT3}	A_{D1}^{DT4}	A_{D1}^{DT5}	A_{D1}^{DT6}	A_{D1}^{DT7}
D2: Degree of digitization	A_{D2}^{DT1}	A_{D2}^{DT2}	A_{D2}^{DT3}	A_{D2}^{DT4}	A_{D2}^{DT5}	A_{D2}^{DT6}	A_{D2}^{DT7}
Value Proposition							
D3: Product type	A_{D3}^{DT1}	A_{D3}^{DT2}	A_{D3}^{DT3}	A_{D3}^{DT4}	A_{D3}^{DT5}	A_{D3}^{DT6}	A_{D3}^{DT7}
D4: Strategy for differentiation	A_{D4}^{DT1}	A_{D4}^{DT2}	A_{D4}^{DT3}	A_{D4}^{DT4}	A_{D4}^{DT5}	A_{D4}^{DT6}	A_{D4}^{DT7}
D5: Target customers	A_{D5}^{DT1}	A_{D5}^{DT2}	A_{D5}^{DT3}	A_{D5}^{DT4}	A_{D5}^{DT5}	A_{D5}^{DT6}	A_{D5}^{DT7}
D6: Value delivery process	A_{D6}^{DT1}	A_{D6}^{DT2}	A_{D6}^{DT3}	A_{D6}^{DT4}	A_{D6}^{DT5}	A_{D6}^{DT6}	A_{D6}^{DT7}
D7: Platform Actors Segments	A_{D7}^{DT1}	A_{D7}^{DT2}	A_{D7}^{DT3}	A_{D7}^{DT4}	A_{D7}^{DT5}	A_{D7}^{DT6}	A_{D7}^{DT7}
Value Creation							
D8: Sourcing	A_{D8}^{DT1}	A_{D8}^{DT2}	A_{D8}^{DT3}	A_{D8}^{DT4}	A_{D8}^{DT5}	A_{D8}^{DT6}	A_{D8}^{DT7}
D9: Third parties involved	A_{D9}^{DT1}	A_{D9}^{DT2}	A_{D9}^{DT3}	A_{D9}^{DT4}	A_{D9}^{DT5}	A_{D9}^{DT6}	A_{D9}^{DT7}
D10: Platform Ecosystem involved	A_{D10}^{DT1}	A_{D10}^{DT2}	A_{D10}^{DT3}	A_{D10}^{DT4}	A_{D10}^{DT5}	A_{D10}^{DT6}	A_{D10}^{DT7}
D11: Value-creation process (key activities)	A_{D11}^{DT1}	A_{D11}^{DT2}	A_{D11}^{DT3}	A_{D11}^{DT4}	A_{D11}^{DT5}	A_{D11}^{DT6}	A_{D11}^{DT7}
D12: Key Resources	A_{D12}^{DT1}	A_{D12}^{DT2}	A_{D12}^{DT3}	A_{D12}^{DT4}	A_{D12}^{DT5}	A_{D12}^{DT6}	A_{D12}^{DT7}
Value Capture							
D13: Revenue model	A_{D13}^{DT1}	A_{D13}^{DT2}	A_{D13}^{DT3}	A_{D13}^{DT4}	A_{D13}^{DT5}	A_{D13}^{DT6}	A_{D13}^{DT7}
D14: Pricing strategy	A_{D14}^{DT1}	A_{D14}^{DT2}	A_{D14}^{DT3}	A_{D14}^{DT4}	A_{D14}^{DT5}	A_{D14}^{DT6}	A_{D14}^{DT7}
D15: Direct profit effect	A_{D15}^{DT1}	A_{D15}^{DT2}	A_{D15}^{DT3}	A_{D15}^{DT4}	A_{D15}^{DT5}	A_{D15}^{DT6}	A_{D15}^{DT7}

The proposed framework could be run through several iterations until all platform patterns are classified and all table cells are filled.

For example, the following sequence based on key dimensions of platform innovation might be implemented:

1. Use of digital platform business model patterns as a first access point for coming up with some initial ideas and a selection of certain patterns for further analysis (see Chapter 1).
2. Reflection on strategic intent, creating business platform layers (Business model and Leadership platform, Talent platform, Delivery platform etc.) from the idea design process while streamlining this with an eye for possible Technology platforms synergies.

Table 2. Matrix form relationship between business model and Information systems platform services dimensions

Business Platform Dimensions	**DT1: Information Systems Platform (ISP)**							
Platform Business Model Pattern Dimensions	**DT11 Employee Digital Collaboration and Workplace**	**DT12 Back-Office Systems**	**DT13 Core Enterprise Systems**	**DT14 Supplier Portal and Apps**	**DT15 Customer Portal and Apps**	**DT16 Business Intelligence**	**DT17 Endpoint Computing**	**DT18 Operational Technology Systems**
Overarching								
D1: Hierarchical impact	A_{D1}^{DT11}	A_{D1}^{DT12}	A_{D1}^{DT13}	A_{D1}^{DT14}	A_{D1}^{DT15}	A_{D1}^{DT6}	A_{D1}^{DT17}	A_{D1}^{DT18}
D2: Degree of digitization	A_{D2}^{DT11}	A_{D2}^{DT12}	A_{D2}^{DT13}	A_{D2}^{DT14}	A_{D2}^{DT15}	A_{D2}^{DT6}	A_{D2}^{DT17}	A_{D2}^{DT18}
Value Proposition								
D3: Product type	A_{D3}^{DT11}	A_{D3}^{DT12}	A_{D3}^{DT13}	A_{D3}^{DT14}	A_{D3}^{DT15}	A_{D3}^{DT6}	A_{D3}^{DT17}	A_{D3}^{DT18}
D4: Strategy for differentiation	A_{D4}^{DT11}	A_{D4}^{DT12}	A_{D4}^{DT13}	A_{D4}^{DT14}	A_{D4}^{DT15}	A_{D4}^{DT6}	A_{D4}^{DT17}	A_{D4}^{DT18}
D5: Target customers	A_{D5}^{DT11}	A_{D5}^{DT12}	A_{D45}^{T13}	A_{D5}^{DT14}	A_{D5}^{DT15}	A_{D5}^{DT6}	A_{D5}^{DT17}	A_{D5}^{DT18}
D6: Value delivery process	A_{D6}^{DT11}	A_{D6}^{DT12}	A_{D6}^{DT13}	A_{D6}^{DT14}	A_{D6}^{DT15}	A_{D6}^{DT6}	A_{D6}^{DT17}	A_{D6}^{DT18}
D7: Platform Actors Segments	A_{D7}^{DT11}	A_{D7}^{DT12}	A_{D7}^{DT13}	A_{D7}^{DT14}	A_{D7}^{DT15}	A_{D7}^{DT6}	A_{D7}^{DT17}	A_{D7}^{DT18}
Value Creation								
D8: Sourcing	A_{D8}^{DT11}	A_{D8}^{DT12}	A_{D8}^{DT13}	A_{D8}^{DT14}	A_{D8}^{DT15}	A_{D8}^{DT6}	A_{D8}^{DT17}	A_{D8}^{DT18}
D9: Third parties involved	A_{D9}^{DT11}	A_{D9}^{DT12}	A_{D9}^{DT13}	A_{D9}^{DT14}	A_{D9}^{DT15}	A_{D9}^{DT6}	A_{D9}^{DT17}	A_{D9}^{DT18}
D10: Platform Ecosystem involved	A_{D10}^{T11}	A_{D10}^{T12}	A_{D10}^{T13}	A_{D10}^{T14}	A_{D10}^{T15}	A_{D10}^{DT6}	A_{D10}^{T17}	A_{D10}^{T18}
D11: Value-creation process (key activities)	A_{D11}^{T11}	A_{D11}^{T12}	A_{D11}^{T13}	A_{D11}^{T14}	A_{D11}^{T15}	A_{D11}^{DT6}	A_{D11}^{T17}	A_{D11}^{T18}
D12: Key Resources	A_{D12}^{T11}	A_{D12}^{T12}	A_{D12}^{T13}	A_{D12}^{T14}	A_{D12}^{T15}	A_{D12}^{DT6}	A_{D12}^{T17}	A_{D12}^{T18}
Value Capture								
D13: Revenue model	A_{D13}^{T11}	A_{D13}^{T12}	A_{D13}^{T13}	A_{D13}^{T14}	A_{D13}^{T15}	A_{D13}^{DT6}	A_{D13}^{T17}	A_{D13}^{T18}
D14: Pricing strategy	A_{D14}^{T11}	A_{D14}^{T12}	A_{D14}^{T13}	A_{D14}^{T14}	A_{D14}^{T15}	A_{D14}^{DT6}	A_{D14}^{T17}	A_{D14}^{T18}
D15: Direct profit effect	A_{D15}^{T11}	A_{D15}^{T12}	A_{D15}^{T13}	A_{D15}^{T14}	A_{D15}^{T15}	A_{D15}^{DT6}	A_{D15}^{T17}	A_{D15}^{T18}

3. Use of value drivers for specific dimensions of the platform innovation to clarify how technology is employed in the implementation of digital platform business model ideas.

Value drivers related to all specific digital dimensions of the platform innovation may be a consideration. This approach keeps the focus on the digital platform transformation throughout the platform business model design process, while providing leeway to also explore opportunities beyond digitalization.

Another example is doing a "checklist" exercise to determine what platform parts are missing in the enterprise, or need to be improved and modernized. The results of such checklist will serve as the foundational elements of the digital business technology roadmap that organizations should build (or update).

CONCLUSION

In reply to the challenges of the digital economy we need more than just tools for the development of a digital technology. As it was mentioned (Blaschke et al., 2017a, b) what we need is "a methodology for systematic Digital Business Modelling based on a language both business and technology experts equally understand". The structured approach we have presented in this chapter is a step in this direction.

REFERENCES

Al-Debei, M., & Avison, D. (2010). Developing a unified framework of the business model concept. *European Journal of Information Systems*, *19*(3), 359–376. doi:10.1057/ejis.2010.21

Blaschke, M., Haki, M., Riss, U., & Aier, S. (2017a). Design Principles for Business-Model-based Management Methods. *A Service-dominant Logic Perspective, Proceedings International Conference on Design Science Research in Information Systems and Technology (DESRIST) 2017.*

Blaschke, M., Cigaina, M., Riss, U., & Shoshan, I. (2017b). Designing Business Models for the Digital Economy. In G. Oswald & M. Kleinemeier (Eds.), *Shaping the Digital Enterprise. Trends and Use Cases in Digital Innovation and Transformation*. Springer International Publishing Switzerland. doi:10.1007/978-3-319-40967-2_6

Blosch, M., & Burton, B. (2016). *Architecting Platform Ecosystems Is Key to Integrate People, Business and Things.* Gartner Research Report, ID: G00291299. Retrieved December 18, 2017, from: https://www.gartner.com/doc/3196533?ref=SiteSearch&sthkw=Platform%20Business%20Model%20Orchestration&fnl=search&srcId=1-3478922254

Boudreau, K. (2010, October). Open Platform Strategies and Innovation: Granting Access vs. Devolving Control. *Management Science, 56*(10), 1849–1872. doi:10.1287/mnsc.1100.1215

Boudreau, K. J., Lacetera, N., & Lakhani, K. R. (2011). Incentives and problem uncertainty in innovation contests: An empirical analysis. *Management Science, 57*(5), 843–863. doi:10.1287/mnsc.1110.1322

Burton, B., & Basiliere, P. (2016). *Architect Digital Platforms to Deliver Business Value and Outcomes.* Gartner Report, ID: G00318861. Retrieved December 18, 2017, from: https://www.gartner.com/doc/3518917?ref=SiteSearch&sthkw=PLATFORM%20STACK&fnl=search&srcId=1-3478922254

Cearley, D., Walker, M., Burke, B., & Searle, S. (2017). *Top 10 Strategic Technology Trends for 2017: A Gartner Trend Insight Report.* Gartner ID: G00319572. Retrieved December 18, 2017, from: https://www.gartner.com/doc/3645332?srcId=1-6595640781

Chesbrough, H., & Rosenbloom, R. (2002). The role of the business model in capturing value from innovation: Evidence from Xerox Corporation's technology spin-off companies. *Industrial and Corporate Change, 11*(3), 529–555. doi:10.1093/icc/11.3.529

Colombo, G., Buganza, T., Klanner, I.-M., & Roiser, S. (2013). Crowdsourcing Intermediaries and problem typologies: An explorative study. *International Journal of Innovation Management, 17*(2), 1–24. doi:10.1142/S1363919613500059

Colombo, G., Dell'Era, C., & Frattini, F. (2015). Exploring the contribution of innovation intermediaries to the new product development (NPD) process: A typology and an empirical study. *R & D Management, 45*(2), 126–146. doi:10.1111/radm.12056

Coltman, T., & Queiroz, M. (2015). Information Technology and Strategy, In M. Augier & D. J. Teece (Eds.), *The Palgrave Encyclopedia of Strategic Management.* Retrieved December 18, 2017, from: http://www.palgraveconnect.com/esm/doifinder/10.1057/9781137294678.0794

Evans, P. C., & Gawer, A. (2016). *The Rise of the Platform Enterprise A Global Survey*. New York: Center for Global Enterprise. Retrieved December 18, 2017, from: http://www.thecge.net/wp-content/uploads/2016/01/PDF-WEBPlatform-Survey_01_12.pdf

Gartner. (2017). *Digitalization. Gartner IT Glossary*. Retrieved December 18, 2018, from: https://www.gartner.com/it-glossary/digitalization/

Guttridge, K., Pezzini, M., Golluscio, E., Thoo, E., Iijima, K., & Wilco, M. (2017). *Magic Quadrant for Enterprise Integration Platform as a Service*. Gartner research ID: G00304070.

Hagiu, A. (2014). Strategic Decisions for Multi-sided Platforms. *MIT Sloan Management Review*, *55*(2).

Hunter, R., & Coleman, M. (2016). Competing for Top Talent: Build the Talent Platform. Gartner Research, ID: G00308714.

IDC. (2017). *Third-platform concept*. Retrieved December 18, 2017, from: http://www.idc.com/promo/thirdplatform/

LeHong, H., Howard, C., Gaughan, D. & Logan, D. (2016). *Building a Digital Business Technology Platform*. Gartner Research ID: G00297286.

Leiblein, M. (2015). Platform Innovation. In M. Augier & D. J. Teece (Eds.), *The Palgrave Encyclopedia of Strategic Management*. Retrieved December 18, 2017, from: http://www.palgraveconnect.com/esm/doifinder/10.1057/9781137294678.0794

Lopez, J. (2014). Digital business is everyone's business. *Forbes*. Retrieved December 18, 2017, from: http://onforb.es/1uBkChg

Lopez, J. (2017). *Digital Business Success Depends on Civilization Infrastructure: A Gartner Trend Insight Report*. ID: G00325088. Retrieved December 18, 2017, from: https://www.gartner.com/doc/3657717?srcId=1-3132930191

Moyer, K. (2016). Industry Visions for Digital Business Set the Terms of Competition. Gartner research, ID: G00307722.

Osterwalder, A., & Pigneur, I. (2010). *Business Model Generation: A Handbook for Visionaries, Game Changers, and Challengers*. John Wiley & Sons.

Osterwalder, A., Pigneur, Y., Bernarda, G., & Smith, A. (2014). *Value Proposition Design: How to Create Products and Services Customers Want.* John Wiley & Sons.

Oswald, G., & Kleinemeier, M. (Eds.). (2017). *Shaping the Digital Enterprise. Trends and Use Cases in Digital Innovation and Transformation.* Springer International Publishing.

Remane, G., Hanelt, A., Tesch, J., & Kolbe, L. (2017). The Business Model Pattern Database – A Tool for Systematic Business Model Innovation. International Journal of Innovation Management, 21(1).

Teece, D. (2012). Next generation competition: New concepts for understanding how innovation shapes competition and policy in the digital economy. *The Journal of Law, Economics & Policy*, *9*, 97–118.

Yablonsky, S. (2016). Intermediaries in E-Commerce. In *Encyclopedia of E-Commerce Development, Implementation, and Management* (pp. 48–73). IGI Global. doi:10.4018/978-1-4666-9787-4.ch005

Yablonsky, S. (2017). Multidimensional cloud-enabled innovations for financial services. *Int. J. Business Excellence*, *11*(4), 464–486. doi:10.1504/IJBEX.2017.082587

Zott, C., Amit, R., & Massa, L. (2010). *The business model: Theoretical roots, recent developments, and future research.* Retrieved December 18, 2017, from: http://www.iese.edu/research/pdfs/DI-0862-E.pdf

Chapter 3
Innovation Platforms:
Data and Analytics Platforms

ABSTRACT

Data and analytics is the heart of a digital business platform. Today, big data (BD) becomes useful when it enriches decision making that is enhanced by application of analytical techniques and some element of human interaction. With the merging of data and information vs. knowledge and intelligence, this chapter investigates an opportunity for cross-fertilization between BD and the field of digital business with related disciplines. Primary BD and analytics platform is a set of business capabilities. This chapter aims to investigate the potential relationship of BD and analytics platform and digital business platform. In doing so, it develops a BD value chain framework, BD business model pattern (BDBMP) with related levels of BD maturity improvement. This framework could be used to find answers on the basic BD and digital business relationship questions.

INTRODUCTION

Over the last decade, the business has become more and more focused on data. This trend is a consequence of the success of many organizations that have used the data they collect to drive their business. The data and analytics platform supports the digital business platform. White and Oestreich (2017) argue that a data and analytics platform is not a technology in its self:

DOI: 10.4018/978-1-5225-5457-8.ch003

Primary it is a set of business capabilities that need to be defined, prioritized, then balanced (investment wise) to sustain the desired maturity level across the broader digital business platform. Once this has been considered and mapped to your digital strategy, maturity and investment in people, process, data and analytics technology need to be adjusted in order to prepare for and assist the transition to digital business.

The emergence of a new wave of data from sources, such as the Internet of Things (IoT), Sensor Networks, Open Data on the Web, data from mobile applications, social network data, together with the natural growth of datasets inside organizations, creates new ways to reuse and extract value from such BD (BD) assets (Manyika et al. 2011; Forbes, 2015; Cavanillas, Curry, Wahlster, 2016; BDVA, 2017).

The ability to effectively manage data, information and extract knowledge is now seen as a key competitive advantage, and many organizations are building their core business on their ability to collect and analyze information to extract business knowledge and insight. BD is related with the emerging digital economy and becomes useful when it enriches decision-making that is enhanced by application of analytical techniques and some elements of human interaction. Digital technologies, including the Internet, company Intranets and other related technologies, are changing the way the companies do business (Peitz, Waldfogel, 2013). BD technology adoption within industrial sectors is an imperative need for most organizations to gain competitive advantage.

BD provides a wealth of potential insight that can transform asset management in ways that help companies get more value from their assets. Nevertheless, conventional maintenance management systems and processes are not enough. Asset managers need ways to turn BD into information, knowledge and insights as well as enable information transparency throughout the entire asset lifecycle and within the ecosystem of operators, equipment manufacturers, service providers etc.

Thus, Internet and related technologies are more than just new digital channels: they are a completely different approach of providing business services and creating new business ecosystems and business models. BD and BD analytics are increasingly seen as a core capability for digital business. This will require organizations to reconsider long-standing process and process modification approaches on the path to intelligent business operations and BD assets management.

The Gartner Group (2017a) defined BD as *3 V's of data*: *high Volume* (amount of data), *high Velocity* (speed of data) and *high Variety* of information assets (range of data types/sources) demanding cost effective, innovative forms of information processing that enable enhanced insight, decision making and process automation. The volume of data is speedily growing. Manyika et al. (2011) terms BD as datasets, which size is beyond the ability of typical database software tools to capture, store, manage, and analyze. Therefore, BD is interpreted as "information assets characterized by such a high volume, velocity and variety [as] to require specific technology and analytical methods for its transformation into value" (De Mauro et al., 2016, p. 103). The variety of BD refers to 30+ different types and sources of data that are available to organizations. By 2020, there will be more than 16 zettabytes of useful data, which implies a growth of 236% per year from 2013 to 2020 (BDVA, 2017). As the BD field have been maturing, other Vs, such as Veracity and Value have been added. *Veracity* brings up the BD accuracy, documenting quality plus uncertainty and reliability of BD while *Value* studies the cost-benefit of BD (Janvrin and Watson, 2017a,b). Now BD requires new forms of data processing to enable enhanced decision-making, insight discovery, and process optimization (Cavanillas, Curry, Wahlster, 2016). As volume, velocity, and variety have increased, along with dropping costs, they have allowed increased analysis of the new compute and storage infrastructures, enabling better strategic, tactical, and operational decision-making (Beyer and Laney, 2012). BD has grown accordingly, bringing new metrics (Manyika, et. al. 2011) and is a major force in all industries (Gartner group, 2013).

The ability to effectively manage BD assets is now seen as the key competitive advantage, and many organizations are building their core business on their ability to collect and analyze information to extract business knowledge and insight from BD. BD technology adoption within industrial sectors is an imperative need for most organizations to gain competitive advantage. This explains why there has been a rapid global transition from industrial economies to knowledge-based or data-driven economies, in which wealth is created by developing and managing intangible assets (Romano et al., 2014, Ricceri and Guthrie, 2009; Guthrie et al., 2001; Intezari and Gressel, 2017; Lerro et al., 2014).

The growth of digitization of economy is exposing traditional asset's limitations, as the boundaries between the technologies and the business blur and new asset classes emerge. This requires a seamless, digital representation of physical assets that allows for the digitalization of key processes. Practitioners as well as decision makers can use this digital representation of assets to easily access all relevant information to conduct business – wherever they are and in a format they can digest quickly. The IoT creates a digital version of physical world devices. This digital representation is the key to the digital transformation of asset management. It allows to reinvent your business models, business processes, and the way people work. Everything is connected in real time. In addition, everyone can access the right information about assets – at just the right time – and collaborate easily with each other because:

- Assets are connected, where needed.
- Asset system behavior is analyzed in real time, simulated, and predicted.
- Asset management information is shared easily in the cloud.

Native digital newcomers are disrupting traditional industries. The digital platforms like Uber, Airbnb, and Facebook, enable intellectual capital and relationships to scale fast and at near-zero cost. Even when these firms rely on physical assets – like property for Airbnb and cars for Uber – they own a participative network platform and a scalable technology, those not being the exact physical asset. They are sharing access to physical assets not owning them. Digital transformation requires that companies reallocate their asset portfolio to support new, digitally enabled business models. But shifting a company's asset portfolio is a lengthy and uncertain process for leaders comfortable with older asset types. That's why existing organizations are confronting with digital transformation now. The main imperatives of asset management have not changed – enterprise still need to balance costs, risk, and performance, currently in accordance with ISO 55001. Other new management standards are already available to mission critical business processes (Krüge, 2016).

Business analytics (BA) and business intelligence (BI) for BD involves collection of a wide variety of inputs, including publicly available data/information/knowledge, human intelligence and active gathering, then processing the resulting inputs to better understand and predict competitor strategies and actions (Erickson and Rothberg, 2015, Marr et al., 2004; Marr, 2015). In a number of ways, BA actually anticipates interest in BD more than other BD disciplines do. BI has always had a broader view of what pieces of data or information might be of use, and BI operations have typically been more focused on analysis of data/information/knowledge assets, looking for insights through correlation and combination. The entire process of identifying, codifying/capturing and sharing BD throughout the firm's network, however, makes competitors that employ BA and BI techniques for BD much more vulnerable. For example, in financial services huge amounts of data are processed to detect issues such as fraud and risk and to analyze customer behavior, segmentation, trading etc. BD analysis and visualization will unlock new use cases and permit new techniques to be put into practice (Yablonsky, 2016a; BDVA, 2017). Capabilities include managing regulation, reporting, audits and compliance, and the automatic detection of behavior patterns and cyberattacks (BDVA, 2017). Open sources of information can be combined with proprietary knowledge to analyze competitive positions, so that recommendation engines will be able to identify potential customers for products (Yablonsky, 2016a; BDVA, 2017).

Reading the environment and discerning the best strategy for BD (full development with minimal investment) and protection from competitors' BA/BI analysis (full set of safeguards to no investment at all) is the challenge of BD initiatives. For example, in accounting the effects of the continual digitization of accounting practice and the accounting profession were researched by Payne (2014) and Quattrone (2016). Accountants tend to focus on the veracity and value benefits of BD. Data volume and velocity have been around in accounting since the 1990s when first Enterprise Resource Planning (ERP) systems were created to handle high volumes of transactions and flow of information insight and of a BD is the bigger variety of data generated externally by a variety of sources including websites, texts, IoTs, social networks and other sources. ERP and other enterprise systems have been collecting financial and non-financial information for decades in the traditional domains of accountants like the general ledger, assurance services and other accounting applications (Janvrin and Watson, 2017a,b). Joia (2001) and Wernke (2002) had already emphasized that despite the difficulty, it is possible for intangible assets (knowledge) to be measured. It highlights that markets, when assessing the actions of some companies based on knowledge, measure it with a value well above that, registered in the accounting books. (Osinski, Selig, Matos, Roman, 2017). Today the assurance services examine and assure a wide variety of different types of data such as systems reliability and e-commerce. Trust Services programs (including Webtrust and Systrust) in USA include examining the security, availability, processing integrity, online privacy and confidentiality of information technology (AICPA, 2016). Janvrin and Watson (2017a,b) argue that expanding assurance skills to new sources of external BD can be seen as the next step in the evolution of accounting. The up to date research has been primarily focused on social media and BD (Arnaboldi, Busco, Cuganesan, 2017). Manning (2015) focused on the network dimension of managing social capital. Wessels (2004) describes the place of information technology in the education of professional accountants. The Journal of Accounting Education dedicated a special issue to the BD (Janvrin and Watson, 2017a,b).

The success of BD usage guides us to ask how we can measure the value of BD. Data is an intangible asset in the same way as patents or intellectual property (Liebowitz, 2013; Ohlhorst, 2013). Certainly, for intellectual property to have value, it has to be recorded and embodied as data. Though organizations experience real costs to create, acquire, store, secure, process and access data, most only track the cost of data storage and loss prevention. This simplistic approach, accounting for the cost of data storage only, suggests that most organizations consider data a commodity, the value of which is entwined in the cost of storage. Basically, many organizations are still valuing data the same as the underlying storage instead of considering the true potential value of data as an asset, distinct from where it is housed.

The general business value of BD is clear (Gartner Group, 2013): it is recognized to improve operational excellence, create a better understanding of customer relations, improve risk management and drive business model innovation. BD in enterprise systems, customer relationship management systems, and similar IT structures/sensors, can provide insights, especially to competitors trying to distinguish the strategic and tactical direction of a firm (Rothberg & Erickson, 2005). This leads directly into advocating all of BD as a potential store of value that can be turned to competitive advantage of the firm. At the same time 25% of large enterprises will have a digital ethics code of conduct, to avoid the abuse of information and ensure consumer value (Gartner Group, 2013). BD in enterprise systems, customer relationship management (CRM) systems and similar IT systems, structures and sensors can provide insights, especially to competitors trying to distinguish the strategic and tactical direction of a firm (Erickson and Rothberg, 2015). This leads directly into advocating all of BD as a potential store of value that can be turned into a competitive advantage of the firm.

This line of thinking belies the true value of data. While storage has fixed cost regardless of whether it contains any data, the data it contains has real value, and storage that contains data has additional cost (Liebowitz, 2013). However, according to the IDC Digital Universe Study in 2013 (IDC, 2014) only 22% of the information in the digital universe was a candidate for analysis and only 5% was actually analyzed. IDC predicts that the valuable percentage will grow to more than 35% by 2020 and more than 10% will be analyzed to produce fully useful insights.

One point present in research on BD, BD assets, and BD development is a clear emphasis on the nature of BD, information and knowledge. A distinction between knowledge, information, and data is quite apparent and very deliberate (Zack, 1999; Erickson, Rothberg, 2014) data, information, knowledge, and wisdom (DIKW) hierarchy. The optimal balance is going to be dependent on circumstances. Reading the environment and discerning the best strategy for BD (full development to minimal investment) and protection from competitors' BA/BI analysis (full set of safeguards to no investment at all) is the challenge of BD initiatives. The entire process of identifying, codifying/capturing, and sharing BD throughout a firm's network, however, makes it much more vulnerable to competitors employing BD business analytics and business intelligence techniques. Business analytics (BA) and business intelligence (BI) involves the collection of a wide variety of inputs, including publicly available data/information/knowledge, human intelligence, and active gathering, then processing the resulting inputs to better understand and predict competitor strategies and actions (Erickson & Rothberg 2014, 2015, Marr, 2015). In a number of ways, business analytics actually anticipates interest in BD more than other BD disciplines do. Business intelligence has always had a broader view of what pieces of data or information might be of use, and BI operations have typically been more focused on analysis of data/information/knowledge assets, looking for insights through correlation and combination.

As we know from knowledge and intelligence approaches, the information and data don't reveal their full value until insights are drawn from them. And so, BD becomes useful when it improves decision-making. Decision-making is enhanced only when analytical techniques are applied and some element of human interaction is applied as well (Erickson, Rothberg, 2014).

A *BD decision intelligence pipeline* (Figure 1) is an adaptation of Gartner Group (2013) pipeline for BD. It represents the sequence of operational steps that have the goal of supporting context-aware BD business decision making. It could be used as a model for a systematic approach to business decision management across a variety of business scenarios. It includes provision for data ingestion, management, conditioning, composition, fixed and adaptive analysis and, finally, injection into the decision process.

Figure 1. A BD decision intelligence pipeline (adopted from Gartner Group, 2013)

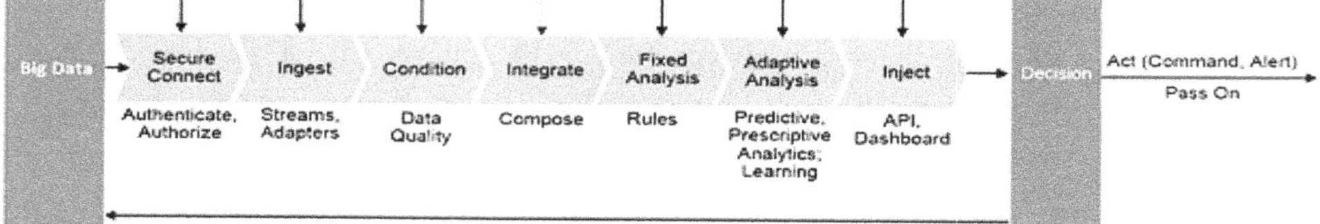

BD Ecosystem and Value Chain

In terms of BD, the ecosystem metaphor is useful to describe the data environment supported by a community of interacting organizations and individuals (Kim et al. 2010; Cavanillas, Curry, Wahlster, 2016). In a business context ecosystem is a network of mutually supportive organizations - including network of internal employees, resources, and activities, and external partners, advisers, customers, suppliers, service providers, and distribution channels (Applegate, Carlson, 2014). BD ecosystems could be formed in different ways around an organization, technology and community platforms. BD ecosystems exist within many industrial sectors where vast amount of data move between actors within complex data/information/knowledge supply chains in different industries, for example, Healthcare, Finance, Logistics, Media, Manufacturing, and Pharmaceuticals (Cavanillas, Curry, Wahlster, 2016). Initiatives like smart cities are showing how different sectors (i.e. energy and transport) can collaborate to maximize the potential for optimization and value return. The cross-fertilization of stakeholder and datasets from different sectors is a key element for advancing the global BD economy (Cavanillas, Curry, Wahlster, 2016), Ecosystems have existed long before, but they were usually called value chains (especially on the micro level) to reflect the linear sequence of value-creating activities involved in designing and producing a product or service, and then marketing, selling, and delivering it to revenue-generating customers (Applegate, Carlson, 2014). BD ecosystem adopted from Cavanillas, Curry, Wahlster (2016) is shown on Figure 2.

New concepts for BD collection, processing, storing, analyzing, handling, visualization and, most importantly, usage are emerging and new BD-enabled strategy and business models are created around them. There are three key ways to generate value for companies along the BD value chain, regardless of sector or domain:

1. Transforming or optimizing and improving the core business;
2. Selling data services; and, perhaps, most importantly,
3. Creating entirely new business models and business development (BDVA, 2017).

Identifying sustainable business models and ecosystems in and across sectors and platforms is an important challenge. In particular, many SMEs that are now involved in highly specific or niche roles will need support to help them align and adapt to new value chain opportunities in the future.

Figure 2. BD ecosystem (adopted from Cavanillas, Curry, Wahlster, 2016)

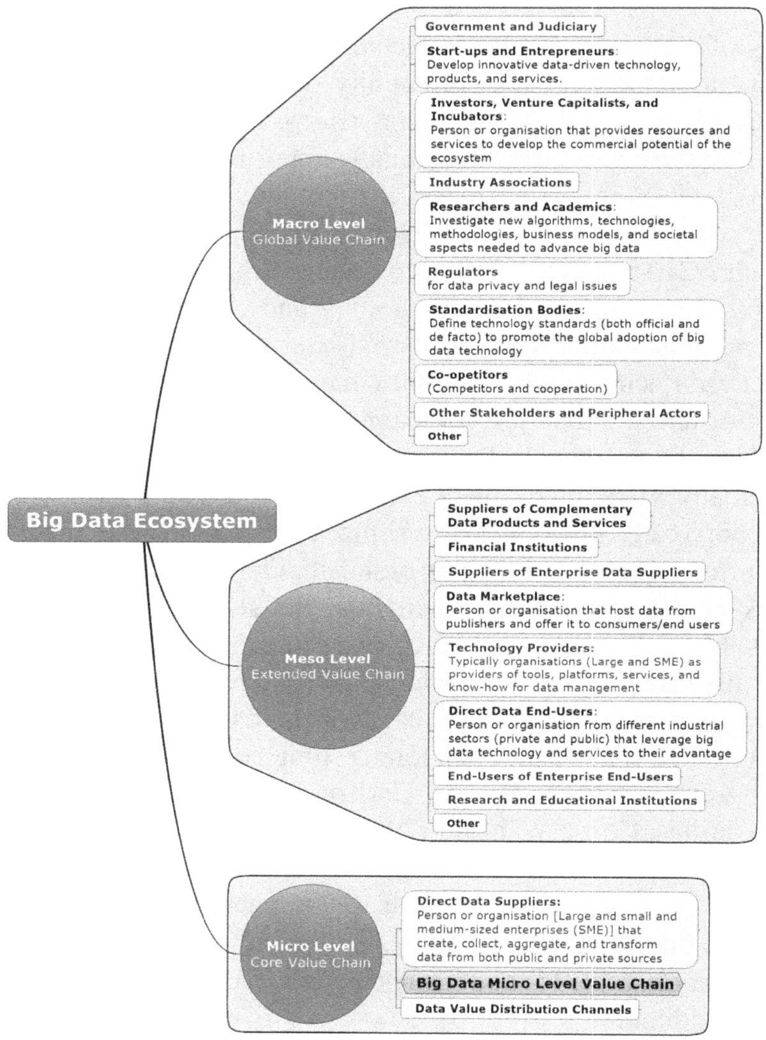

The real value of BD could be determined through the life cycle of BD. We try to plot the value of BD over its life cycle as a framework to consider how an organization might determine the value of enterprise data (Liebowitz, 2013). Curry (2016) defines Data Value Chain as follows:

A value chain is made up of a series of subsystems each with inputs, transformation processes, and outputs… As an analytical tool, the value chain can be applied to information flows to understand the value creation of data technology. In a Data Value Chain, information flow is described as a series of steps needed to generate value and useful insights from data. The European Commission sees the data value chain as the "centre of the future knowledge economy, bringing the opportunities of the digital developments to the more traditional sectors (e.g. transport, financial services, health, manufacturing, retail). (p. 31)

We will focus mostly on the BD micro-level value chain (Cavanillas et al., 2016) defined by Curry (2016). The BD micro level value chain is used to model the high-level activities that comprise an enterprise. The BD Micro Level Value Chain, as illustrated in Figure 3, is used to model the high-level activities that comprise an enterprise.

The proposed BD Value Chain identifies the following key high-level BD activities/dimensions:

1. *BD Acquisition* mean the availability of BD and access to BD sources. There is a variety of BD types and sources. Value is created by acquiring data and combining data from different sources. BD pre-processing, validating and augmenting BD, as well as ensuring their integrity and accuracy add value.
2. *BD Analysis* is referred to using the raw data acquired in decision-making as well as domain-specific usage, and providing access to data with low latency while ensuring data integrity and preserving privacy. Evaluation, machine learning, information extraction and data discovery of intangible BD assets add value.
3. *BD Curation* is based on the active management and measurement of BD assets over its life cycle to ensure it meets the necessary BD quality requirements for its effective usage.
4. *Big Data Storage* is the persistence and management of data in a scalable way that satisfies the needs of applications that require fast access to the data. Historically, during the last 30-40 years Relational Database Management Systems (RDBMS) have been the main solution to the storage paradigm for structured data. However, they have some limitations for storing and processing semi/unstructured BD. That is why NoSQL and other new innovative database technologies have been designed.
5. *BD Usage* covers the main BD assets usage in business decision-making that can improve competitiveness through reduction of costs, increased added value, or any other parameter that can be measured against existing performance criteria.

Figure 3. BD Micro Level Value Chain
(adopted and enriched from Cavanillas et al. 2016)

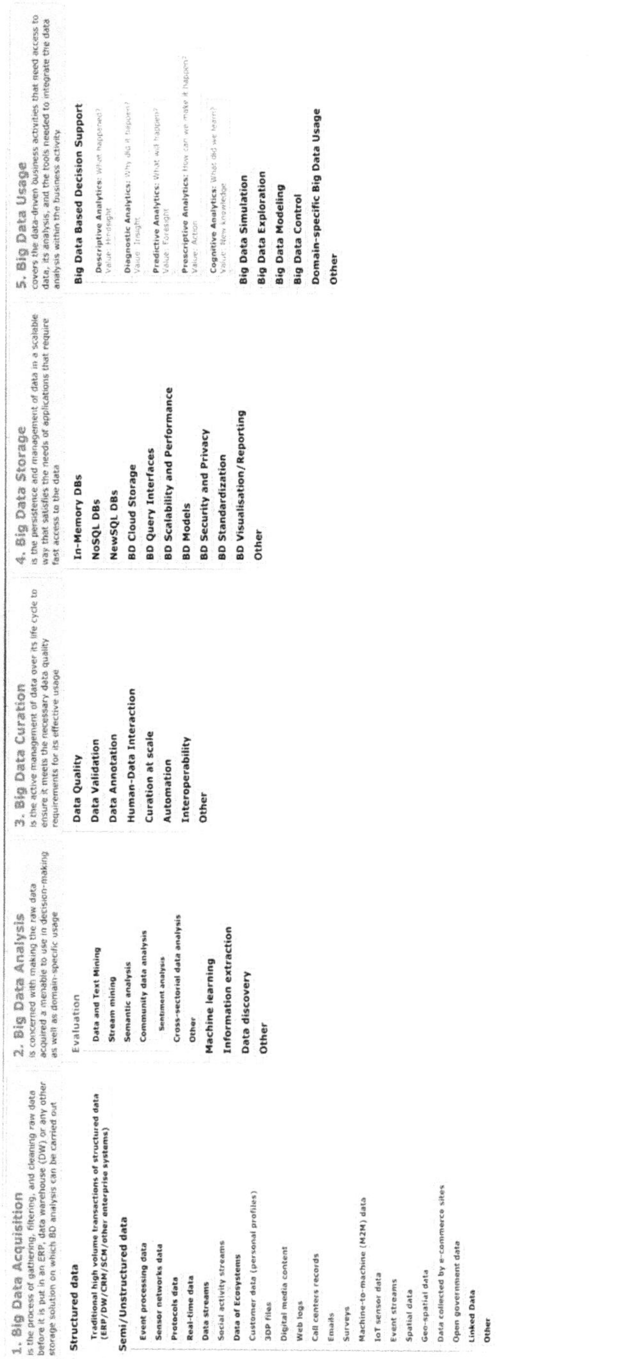

BD platform ecosystems exist within many industrial sectors where vast amount of data move between actors within complex data/information/knowledge supply chains in different industries, for example, Healthcare, Finance, Logistics, Media, Manufacturing and Pharmaceuticals (BDVA, 2017). Initiatives like smart cities are showing how different sectors (i.e. energy and transport) can collaborate to maximize the potential for optimization and value return. The cross-fertilization of stakeholders and datasets from different sectors is a key element for advancing the global BD economy (Cavanillas, Curry and Wahlster, 2016).

The BD value ecosystem will comprise many new stakeholders in the future. The following key stakeholders are seen as the main actors along the BD value chain (Cavanillas, Curry and Wahlster, 2016; BDVA, 2017):

- *BD User Enterprises* in all sectors, who want to improve/transform their services and products using BD technology, data products and services;
- *BD Generators and Providers*, who create, collect, aggregate, transform and model raw data from various public and non-public sources and offer them to customers.
- *Technology Providers*, who provide tools and platforms that offer BD management and analytics tools to extract knowledge from BD, curate and visualise it;
- *BD Service Providers*, who develop BD applications on top of the tools and platforms to provide services to BD User Enterprises.
- *BD Entrepreneurs*, who build innovative data businesses and data services based on BD on the demand and supply side;
- *BD Researchers and academics*, who provide access to state-of-the-art research in Big Data technology;
- *BD Policy makers*, responsible for establishing policy framework conditions that foster the adoption of BD technology in various sectors.

In addition, the following organizations and communities have an impact on BD-driven ecosystems building on the BD value chain (BDVA, 2017):

- *Regulatory bodies* that define privacy and legacy issues related to data usage;
- International/national standardisation bodies that promote new concepts, systems and solutions for global adoption in international standards;

- *Collaborative networks*, where different players in the value chain collaborate to offer value services to their customers based on data value creation.

Creating a favorable ecosystem for BD and promoting its accelerated adoption requires an interdisciplinary approach that addresses main dimensions of BD value (BDVA, 2017).

BD GOVERNANCE FRAMEWORK

According to the Business Analysis Body of Knowledge (Babok, 2015), business analysis is the set of tasks and techniques used to work as a liaison among stakeholders in order to understand the structure, policies, and operations of an organization, and to recommend solutions that enable the organization to achieve its goals. The goal of BD governance is to improve a company's revenue and services or to reduce its costs using new BD capabilities/tools and their relationship with intellectual capital knowledge management. The process of BD governance involves understanding how and what the solution needs to do, what it should look like, and the people or systems that interact with it. The governance process requires to grasp, where the company is today and where it needs to be in order to achieve the goal (Babok, 2015). We identify the activities that enable the company to define the business problem or opportunity in terms of concept modelling using knowledge modelling tools. A BD governance concept modelling is used to organize the business vocabulary needed to consistently and thoroughly communicate the knowledge of a domain. Concept models put a premium on high-quality, design-independent definitions that are free of data or implementation biases. Concept models also emphasize rich vocabulary (Babok, 2015). A concept model identifies the correct choice of terms to use in communications, including all business analysis information. It is especially important where high precision and subtle distinctions need to be made.

Corresponding to data governance framework adopted from the second edition of DAMA International's Guide to the Data Management Body of Knowledge (DAMA-DMBOK2, 2017) we consider BD governance as the exercise of authority and control (planning, monitoring, and enforcement) over the management of BD assets. The BD governance function guides how all other data management functions are performed. BD governance is high-level, executive data stewardship.

Figure 4. A BD governance framework (first level)

Next we have investigated the specifics of BD governance applying data governance framework (DAMA-DMBOK2, 2017) to such main BD knowledge areas (see Figure 4):

- BD Strategy and Business models.
- BD Modeling and Design.
- BD Architecture Management.
- BD Storage and Operations.
- BD Security.
- BD Integration.
- BD BI and Analytics.
- Reference and Master Data for BD.
- Meta-data for BD.
- BD Quality.
- Other.

Each knowledge area is the collection, integration, and presentation of data to knowledge workers for the purpose of BD governance and decision-making. They are composed of activities supporting all phases of the decision support life cycle that provides context, moves and transforms BD from sources to a common target data store, and then provides knowledge workers various means of access, manipulation, and reporting of the integrated target data.

BD governance is a subset of different features and could be presented by taxonomy that includes data, compute and storage infrastructures, analytics, visualization, security/privacy and industry domains, grouped in such subclasses:

- BD management suppliers.
- BD management inputs.

Figure 5. BD Governance Framework partly expanded for knowledge area ¨BD Strategy and Business Models¨ (three levels)

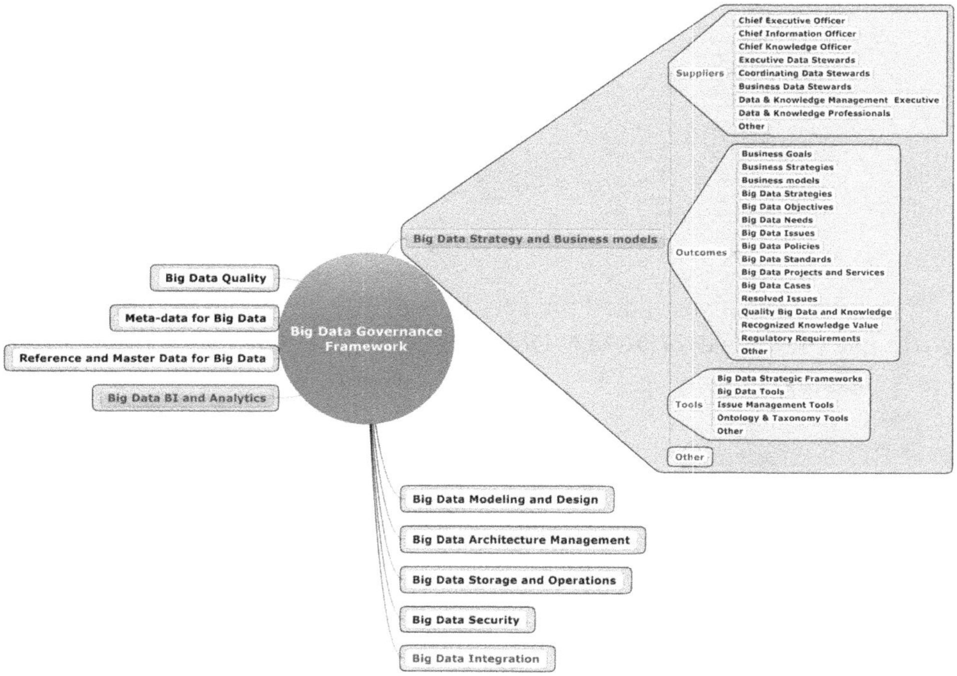

- BD management participants.
- BD management tools.
- BD management primary deliverables.
- BD management consumers.
- BD management metrics.
- BD management product/services classes.
- BD management standards and rules.
- Other.

Based on theoretical conceptualization, combined with empirical evidence, we propose a framework for BD governance (see Figures 4, 5, 6). We have developed five-level taxonomy with more than 100 entities. During the exploratory and descriptive phases of the work secondary sources were mainly used. These are books, scientific and professional articles in journals, surveys, and information from professional online communities. Primary sources were used as well. Qualitative empirical study was conducted in

Figure 6. A BD Governance Framework partly expanded for knowledge area ¨BD BI and Analytics¨ (three levels)

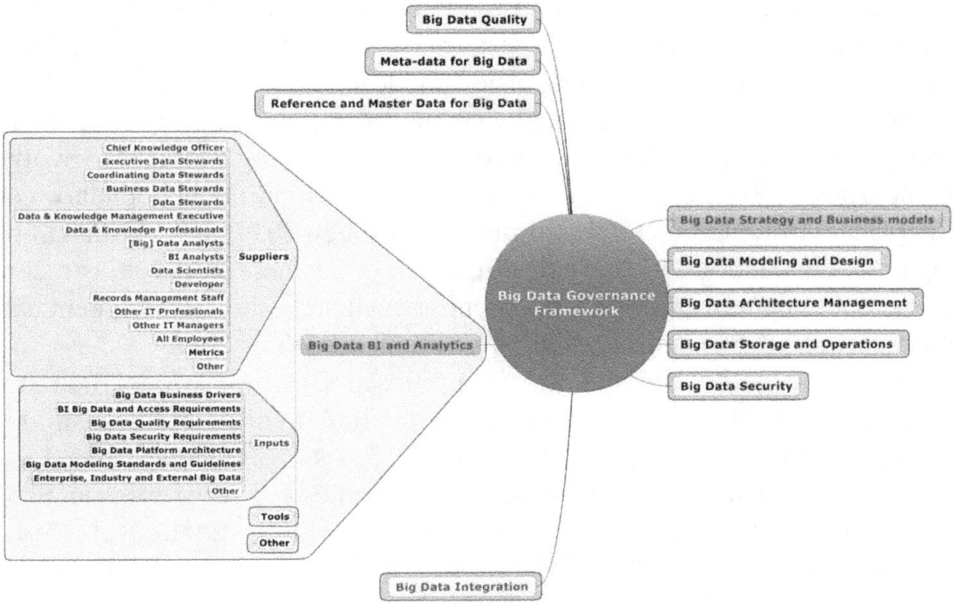

order to acquire relevant data to get opinions of companies' top management from industries that might potentially use and/or be interested in providing platform for consulting innovation services.

As the approach implies we chose to use the semi-structured interviews to get in-depth answers from our respondents.

After the framework had been applied for relevant firms, the results enabled an evaluation of the appropriateness of the framework. Required modifications were identified and a refined framework is proposed. We argue that the "BD Strategy and Business models", and ¨BD BI and Analytics¨ taxonomy concepts are mostly related with BD research area.

BD STRATEGY AND BUSINESS MODELS

Marr (2015) categorizes several parts of a BD strategy which for organizational change and employee engagement purposes will be less centered around customer and competition/risk and more likely to be centered around purpose (the context of the change that is required); finance (the related costs of change, recruitment, development and retention); operations (resource allocation, roles, responsibilities and core competencies to deliver on execution capability); resources (the required systems, infrastructure, processes and data to run business operations; people management, performance, retention and organizational culture; values, strategy, mission, vision and leadership competencies).

Since BD businesses additionally create and capture value by facilitating ecosystems external to the firm, approaches to evaluating BM need to evolve to incorporate BD measures that extend beyond the traditional boundaries of the firm. Gartner Group (2016) argues that through 2017, fewer than half of lagging organizations will have made cultural or business model adjustments sufficient to benefit from BD. We propose evaluating the measurement and management of BD in the context of business models.

Osterwalder and Pigneur (2010) proposed Business Model Patterns (henceforth, BMPs) - business models with similar characteristics, similar arrangements of business model building blocks, or similar behaviors. BMPs help to understand business model dynamics and serve as a source of inspiration for creating e-commerce business models (Gassmann et al. 2014).

Drawing on existing knowledge in the field of business model patterns (Osterwalder and Pigneur, 2010; Blaschke et al., 2017 a,b) and platform research (Evans, and Gawer, 2016) we argue that a meta-structuring perspective serving through the BD platform landscape represents an important strategic platform contribution that remains missing. For that we have to define a meta characteristics of the BD dimensions. The BD business model is a useful lens for understanding a company's underlying logic because it explains what BD value is provided, how this value is created and delivered, and how profits can be generated therefrom (see Figure 7). This concept helps to understand capturing value from BD technological innovations and business models (Osterwalder and Pigneur, 2010), the boundaries of a firm and creating a direct connection between business strategy and business processes. Therefore we replace *basic BD dimensions* with *BD business model pattern dimensions* based on expanded results of Remane et al. (2017) (see Figure 7).

Business Model Patterns concept is another way of structuring digital value drivers (Gassmann et al. 2014). The business model patterns highlight a number of specific design ideas, and they detail them using companies having successfully implemented them as examples. Following Osterwalder (2004) we provide a BD business model pattern framework, which is a conceptualization and formalization of the elements, relationships, vocabulary, and semantics of a BD BM initiatives (Yablonsky, 2016a,b), see Figure 7. BD Business Model Pattern (BDBMP) dimensions framework with meta-components could be used in research of BD identification/creation/assessment/disclosure.

Figure 7. BD business model pattern dimensions framework with meta-components

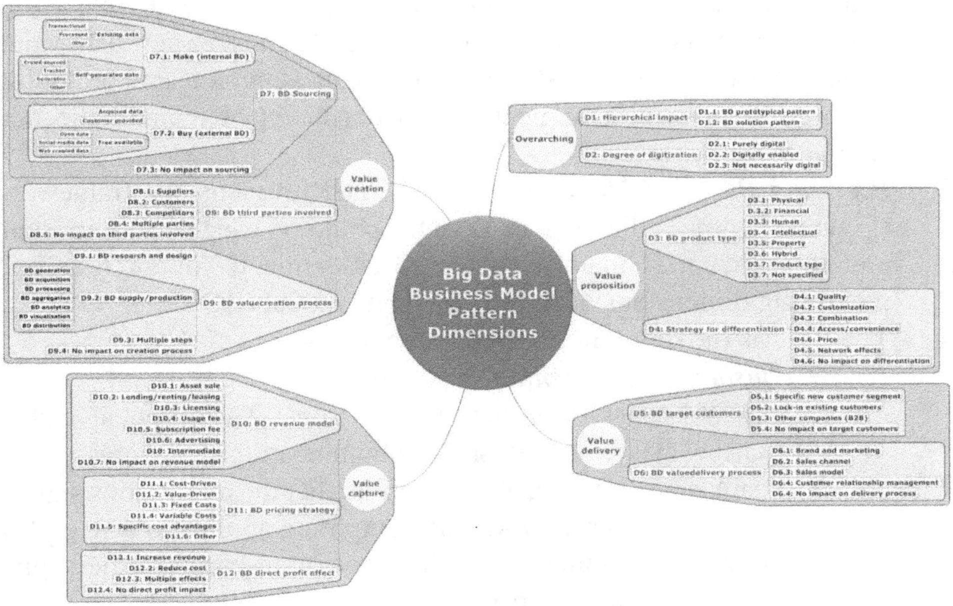

CONCLUSION

The chapter provides a broad overview of BD concept modelling using data management tasks. In spite of the fast growth of the modern information and communication technologies, the importance of BD conceptualization and governance is only growing. Knowledge/information/data management is an important condition for firms to acquire BD and external knowledge for the firms in the high technology industries that are characterized by the high level of uncertainty and rapid rate of change. This research adds the following results to current understanding:

1. *BD decision intelligence pipeline* (Figure 1) represents the sequence of operational steps that have the goal of supporting context-aware BD business decision making. It could be used as a model for a systematic approach to business decision management across a variety of business scenarios. It includes provision for data ingestion, management, conditioning, composition, fixed and adaptive analysis and, finally, injection into the decision process.
2. *BD value chain framework* (Figure 2) proposes a BD life cycle of the operational steps sequence that has the goal of supporting context-aware

BD business decision making has been associated with full cycle of BD maturity model improvement. In a digital age, when intangible assets are becoming more crucial for the firm performance, new BD value dimensions can support CEOs and their management teams providing a more effective measurement and management of their intellectual capital assets.

3. *Big Data Business Model Pattern (BDBMP)* dimensions framework (Figure 7) with 12 dimensions D1-D12 could be used as an important counterpart in BD maturity/structure assessments for analysis and measuring dimensions of the BD maturity model (BDMM).
4. The BD has the following related components:
   a. *BD business instrumentation comprising.*
      i. Firm Ecosystem's New Connections: The physical and/or virtual means of ecosystem communicating or sensing through capturing data from physical infrastructure, information signals and business processes and computing applications (e.g., IoT devices, consumer wearables, ecosystem-based partners) plus social network connections.
      ii. BD Data/Information/Knowledge Exchanges/Sharing: Representations of events and activities that have business meaning within the business infrastructure and are about the Big Data structure and format, as well as raw information.
      iii. BD mining could provide new insides (Wixom, 2013; Wixom et al., 2017).
   b. *Business administration (for non-transactional-centered work).*
      i. BD BI/BA could be used for smart performance management and smart reporting.
      ii. BD business processes could be related to smart business planning and budgeting.
      iii. BD and BD applications could be used in modelling formal but low-variable-based outcomes, core to business administration.
   c. *Intelligent business processes (for transactional-centered work).*
      i. BD decision management and analytics technologies could be exploited to derive capture and convey of firm operational context across one or more interactions executing in a process instance. The decision option selected by the decision model could be acted on by a human or system to dynamically alter the course of the process instance as it executes.

  ii. BD and BD applications could be utilized for more complex and high-variability-based firm outcomes and decisions.
5. BD influences all parts of the firm, transforms business models and ecosystems, indorses new metrics, skills, actors and activities.

As we has explored in this chapter, BD provides a wealth of potential insight that can transform asset management in ways that can help companies get more value from their BD assets. Firms have to focus on continuous auditing, continuous control monitoring and advanced analytics based on BD (AICPA, 2015). Nevertheless, to harness it, conventional maintenance management systems and processes are not enough. Asset managers need ways to turn BD into information, knowledge and insights, as well as enable information transparency throughout the entire BD assets lifecycle and within the BD ecosystem of stakeholders.

Currently, in the FinTech industry huge amounts of data are processed to detect issues such as fraud and risk and to analyze customer behavior, segmentation, trading, etc. BD analysis and visualization will open up new use cases and permit new techniques to be put into practice. Possibilities include managing regulation, reporting, audits and compliance, together with the automatic detection of behavior patterns and cyberattacks. Open sources of information can be combined with proprietary knowledge to analyze competitive positions, so that recommendation engines are be able to identify potential customers for products (Yablonsky, 2016b; BDVA, 2017).

REFERENCES

AICPA. (2015). *Audit Analytics and Continuous Audit: Looking Toward the Future*. American Institute of Certified Public Accountants, Inc. Retrieved December 18, 2017, from: https://www.aicpa.org/InterestAreas/FRC/AssuranceAdvisoryServices/DownloadableDocuments/AuditAnalytics_LookingTowardFuture.pdf

Applegate, L., & Carlson, C. (2014). Entrepreneurship Reading: Attracting Talent and Building Ecosystems. HBS Core Curriculum Readings, # 8068-HTM-ENG.

Arnaboldi, M., Busco, C., & Cuganesan, S. (2017). Accounting, accountability, social media and big data: Revolution or hype? *Accounting, Auditing & Accountability Journal*, *30*(4), 762–776. doi:10.1108/AAAJ-03-2017-2880

Babok. (2015). *BABOK: A Guide to the Business Analysis Body of Knowledge, v3*. Institute of Business Analysis, International (IIBA).

BDVA. (2017). *European Big Data Value Strategic Research and Innovation Agenda, BDVA, version 4.0.* Retrieved December 18, 2017, from: http://www.bdva.eu/sites/default/files/BDVA_SRIA_v4_Ed1.1.pdf

Beyer, M., & Laney, D. (2012). *The Importance of 'Big data': A Definition.* Gartner research report, ID: G00235055. Retrieved December 18, 2017, from: https://www.gartner.com/doc/2057415/importance-big-data-definition

Blaschke, M., Haki, M., Riss, U., & Aier, S. (2017a). Design Principles for Business-Model-based Management Methods—A Service-dominant Logic Perspective. *Proceedings International Conference on Design Science Research in Information Systems and Technology (DESRIST) 2017.*

Blaschke, M., Cigaina, M., Riss, U., & Shoshan, I. (2017b). Designing Business Models for the Digital Economy. In G. Oswald & M. Kleinemeier (Eds.), *Shaping the Digital Enterprise.* Springer International Publishing Switzerland. doi:10.1007/978-3-319-40967-2_6

Cavanillas, J., Curry, E., & Wahlster, W. (Eds.). (2016). *New Horizons for a Data-Driven Economy A Roadmap for Usage and Exploitation of Big Data in Europe*. Springer Open. doi:10.1007/978-3-319-21569-3

DAMA - DMBOK 2. (2017). *The second edition of DAMA International's Guide to the Data Management Body of Knowledge.* DAMA International.

De Mauro, A., Greco, M., & Grimaldi, M. (2016). A formal definition of Big Data based on its essential features. *Library Review*, *65*(3), 122–135. doi:10.1108/LR-06-2015-0061

Erickson, G., & Rothberg, H. (2014). Big Data and Knowledge Management: Establishing a Conceptual Foundation. *Electronic Journal of Knowledge Management*, *12*(2), 108–116. Retrieved from www.ejkm.com

Erickson, G., & Rothberg, H. (2015). Knowledge assets in services across industries and across time. *Investigaciones Europeas de Dirección y Economía de la Empresa*, *21*(2), 58–64. doi:10.1016/j.iedee.2014.09.002

Evans, P. C., & Gawer, A. (2016). *The Rise of the Platform Enterprise A Global Survey*. New York: Center for Global Enterprise. Retrieved from: http://www.thecge.net/wp-content/uploads/2016/01/PDF-WEBPlatform-Survey_01_12.pdf

Forbes. (2015). *Betting on Big data*. Forbes Insights in Cooperation with Teradata.

Gartner. (2017a). *Big Data definition*. Retrieved December 18, 2017, from: https://www.gartner.com/it-glossary/big-data

Gartner Group. (2013). Toolkit: Big Data Business Opportunities From Over 100 Use Cases. Gartner Research report, ID: G00252112.

Gassmann, O., Frankenberger, K., & Csik, M. (2014). *The Business Model Navigator: 55 Models That Will Revolutionise Your Business*. Pearson Education Limited.

Guthrie, J., Petty, R., & Johanson, U. (2001). Sunrise in the knowledge economy: Managing, measuring and reporting intellectual capital. *Accounting, Auditing & Accountability Journal*, *14*(4), 365–384. doi:10.1108/EUM0000000005869

IDC. (2014). *IDC The Digital Universe Study*. IDC.

Intezari, A., & Gressel, S. (2017). Information and reformation in KM systems: Big Data and strategic decision-making. *Journal of Knowledge Management*, *21*(1), 71–91. doi:10.1108/JKM-07-2015-0293

Intezari, A., & Gressel, S. (2017). Information and reformation in KM systems: Big data and strategic decision-making. *Journal of Knowledge Management*, *21*(1), 71–91. doi:10.1108/JKM-07-2015-0293

Janvrin, D., & Watson, M. (Eds.). (2017a). *Journal of Accounting Education*, *38*.

Janvrin, D., & Watson, M. (2017b). "Big Data": A new twist to accounting. *Journal of Accounting Education*, *38*, 3–8. doi:10.1016/j.jaccedu.2016.12.009

Joia, L. A. (2001). Medindo o capital intellectual. *RAE – Revista de Administração de Empresas*, *41*(2), 54-63. Available at: www.scielo.br/pdf/rae/v41n2/v41n2a06

Kim, H., Lee, J.-N., & Han, J. (2010). The role of IT in business ecosystems. *Communications of the ACM*, *53*(5), 151. doi:10.1145/1735223.1735260

Krüge, A. (2016). *Digital Transformation of Asset Management*. SAP paper. Available at: https://www.sap.com/documents/2016/12/0421b662-9f7c-0010-82c7-eda71af511fa.html

Lerro, A., Linzalone, R., & Schiuma, G. (2014). Managing intellectual capital dimensions for organizational value creation. *Journal of Intellectual Capital*, *15*(3), 350–361. doi:10.1108/JIC-05-2014-0063

Liebowitz, J. (Ed.). (2013). *Big Data and Business Analytics*. Auerbach Publications. doi:10.1201/b14700

Manning, P. (2015). Managing Social Capital – The Network Dimension. In The Human Factor In Social Capital Management: The Owner-manager Perspective. Emerald Group Publishing Limited. doi:10.1108/S2043-905920150000009005

Manyika, J., Chui, M., Brown, B., Bughin, J., Dobbs, R., Roxburgh, C., & Byers, A. H. (2011). *Big Data: The next frontier for innovation, competition, and productivity*. McKinsey Global Institute. Retrieved December 18, 2017, from: https://bigdatawg.nist.gov/pdf/MGI_big_data_full_report.pdf

Marr, B. (2015). *Big data: Using SMART Big data, Analytics and Metrics To Make Better Decisions and Improve Performance*. John Wiley and Sons.

Marr, B., Schiuma, G., & Neely, A. (2004). Intellectual capital-defining key performance indicators for organizational knowledge assets. *Business Process Management Journal*, *10*(5), 551–569. doi:10.1108/14637150410559225

Ohlhorst, F. (2013). *Big Data Analytics: Turning Big Data into Big Money*. John Wiley and Sons.

Osinski, M., Selig, P., Matos, F., & Roman, D. (2017). Methods of evaluation of intangible assets and intellectual capital. *Journal of Intellectual Capital*, *18*(3), 470–485. doi:10.1108/JIC-12-2016-0138

Osterwalder, A. (2004). *The Business Model Ontology - A proposition in a design science approach* (Ph.D. thesis). University of Lausanne, Ecole des Hautes Etudes Commerciales HEC.

Osterwalder, A., & Pigneur, I. (2010). *Business Model Generation: A Handbook for Visionaries, Game Changers, and Challengers*. John Wiley & Sons.

Payne, R. (2014). Discussion of 'digitisation, Big Data and the transformation of accounting information' by Alnoor Bhimani and Leslie Willcocks. *Accounting and Business Research*, *44*(4), 491–495. doi:10.1080/00014788.2014.910053

Peitz, M., & Waldfogel, J. (2013), The Oxford Handbook of the Digital Economy. Oxford University Press.

Quattrone, P. (2016). Management accounting goes digital: Will the move make it wiser? *Management Accounting Research*, *31*, 118–122. doi:10.1016/j.mar.2016.01.003

Remane, G., Hanelt, A., Tesch, J., & Kolbe, L. (2017). The Business Model Pattern Database – A Tool for Systematic Business Model Innovation. International Journal of Innovation Management, 21(1).

Ricceri, F., & Guthrie, J. (2009). Critical analysis of international guidelines for the management of knowledge resources. In D. Jemielniak & J. Kociatkiewicz (Eds.), *Handbook on Research on Knowledge Intensive Organisations* (pp. 375–392). Hershey, PA: Information Science Press. doi:10.4018/978-1-60566-176-6.ch023

Romano, A., Passiante, G., Del Vecchio, P., & Secundo, G. (2014). The innovation ecosystem as booster for the innovative entrepreneurship in the smart specialization strategy. *International Journal of Knowledge Based Development, 5*(3), 271–288. doi:10.1504/IJKBD.2014.065315

Wernke, R. (2002). *Identificação de potenciais geradores de intangíveis* (PhD thesis). Universidade Federal de Santa Catarina, Centro Tecnológico. Programa de Pós-Graduação em Engenharia de Produção. Florianópolis-SC, 2002. Retrieved December 18, 2017, from: https://repositorio.ufsc.br/bitstream/handle/123456789/84208/189150.pdf?sequence=1

Wessels, P. L. (2004). Information technology and the education of professional accountants. *Meditari Accountancy Research, 12*(1), 219–234. doi:10.1108/10222529200400012

White, A., & Oestreich, T. (2017). *Data and Analytics Programs Primer for 2017*. Gartner research ID: G00318601.

Wixom, B., & Ross, J. (2017). How to Monetize Your Data. *MIT Sloan Management Review, 58*(3), 10–13.

Wixom, B., Ross, J., Beath, C., & Miller, C. (2013). Capturing Value from Big Data At comScore Through Platform, People, and Perception. *MIT Sloan CISR Research Briefing, 13*(11).

Yablonsky, S. (2016a). Intermediaries in E-Commerce. In I. Lee (Ed.), *Encyclopedia of E-Commerce Development, Implementation, and Management*. IGI Global. doi:10.4018/978-1-4666-9787-4.ch005

Yablonsky, S. (2016b). E-finance services in Russia. In I. Lee (Ed.), Encyclopedia of E-Commerce Development, Implementation, and Management (pp. 1215–1228). IGI Global. doi:10.4018/978-1-4666-9787-4.ch087

Chapter 4
Innovation Platforms:
IOT Platforms

ABSTRACT

This chapter reports on a study of the taxonomy of innovations for the internet of things (IoT) products and services. The analysis and discussion lead to a multidimensional framework of innovations, with a particular emphasis on a technology stack, business models, products, services, and platform innovations. The authors argue that a multi-sided platform might possibly be a successful business model for the adoption of IOT products and services. They develop IOT business-model pattern and the general definition of a IOT multi-sided platform. The research contains guidelines to help practitioners and policy makers develop platform-enabled IoT innovation strategies through the consideration of various levels of business models and MSP strategy. It offers a relevant source of ideas and guidance for anyone interested in research and practice related to rethinking IoT, product, and service innovation.

INTRODUCTION

The IoT device mesh is dynamic and pervasive. Gartner (2014) forecasts that the IoT will reach 26 billion units by 2020, up from 0.9 billion in 2009, and will impact the information available to supply-chain partners and how the supply chain operates. All these billions of things connected with businesses and 3 billion people with smartphones, smart watches, etc. generate vast amounts of rich data, and what companies do with that data – how they

DOI: 10.4018/978-1-5225-5457-8.ch004

turn it into proprietary algorithms and business value – will determine how well they maximize the opportunity presented by digital business. In order to unlock the potential of IoT, organizations have to master new business models, technology architectures, operating systems, tools, methodologies, databases, networks, middleware and sourcing partners. Manufacturers can now begin to define their target manufacturing model and then plan the IoT transformation roadmap.

The Internet of Things has been defined as "a global infrastructure for the information society, enabling advanced services by interconnecting (physical and virtual) things based on existing and evolving interoperable information and communication technologies" (Y.2060 2012). These physical and virtual things or smart things could include any object or smart device, such as clothes, a watch, a suitcase, an animal, or even a person. Wearable technology devices, or simply wearables, refer to smart things that are designed to be comfortably worn on the human body. LexInnova (2015) defines: "Wearable technology tends to provide sensory and scanning capability, such as bio-feedback and tracking of physiological function. Wearables also have communication capability which allows humans to access data in real-time using another connected device or medium. Rapidly evolving examples of wearable devices include smart watches, intelligent eyewear, bio-sensing contact lenses, e-clothing and smart jewelry, such as rings, bracelets, as well as hearing aid-like devices that are designed to look like ear rings. In some cases, wearable devices may also be implanted into the human body."

Wearable devices and their supportive complementary services have been undergoing anticipation, hype and speculation mostly during the last decade (Gartner 2015, 2016a, b, c). IoT has the potential to transform the fashion industry and the way we live and work (Gartner 2015). The wearable smart-devices industry is booming and is being promulgated by the market recently as an alternative to the post-smart-phone industry. It also has unique characteristics of an industrial ecosystem different from the ecosystem of the smart-phone industry.

We analyze the characteristics of wearable applications for IoT scenarios of smart wearables and related products like e-textile, smartwatches (e.g., Samsung's Galaxy Gear, Pebble Watch and Qualcomm's TOQ), wristbands (Polar Loop, Nike Fuel band), wearable bio-monitors, etc. Long-term engagement and consumer adoption in the wearable-technology (henceforth, WT) industry are not yet entirely established. Additionally, several companies strive to understand this field of technology in order to create strategic competitive advantage and innovative business models, as well as sustainable

Figure 1. Innovation platforms

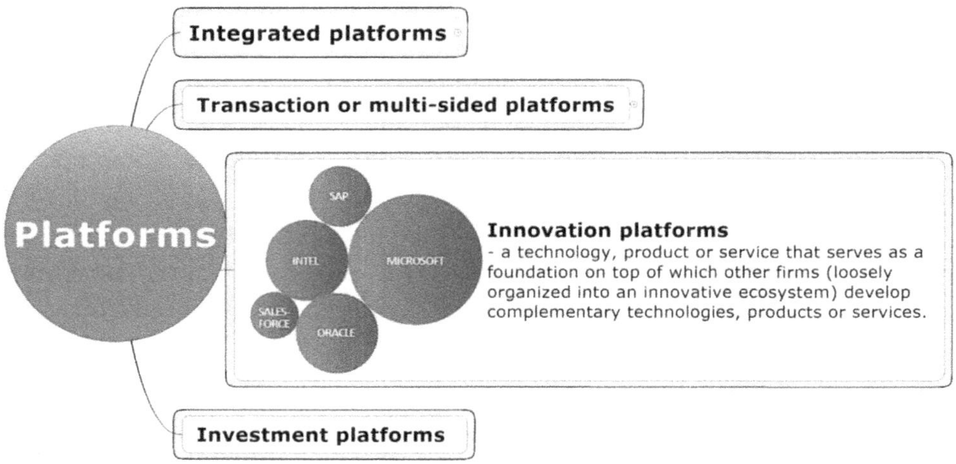

and unique value. Much attention is drawn to the efforts by large companies such as IBM, Microsoft, Google, Cisco, GE, Bosch and government initiatives like Industry 4.0 in Germany (IBM 2015; Microsoft 2015; Google 2015; Cisco 2015; Bosch 2015; Industrie 2013).

IoT platforms belong to Innovation platform type group (see Figure 1).

Our goal is to conceptualize the innovation IoT platforms and wearable-products technology stack with a particular emphasis on business models, products, services and platform innovations. We illustrate, analyze and classify such conceptualization by means of ontologies, in particular lightweight ontologies –taxonomies (Poli et al. 2010; Yablonsky 2014, 2016a, b).

IoT Taxonomy

The IoT and related wearables form one of the most data-rich environments on the Web today. An increasing number of large, midsize and small vendors have been entering the market, ranging from IT megavendors to midsize providers incorporating IoT capabilities into existing product and service offerings (see Figure 2). A clear and precise description and structuring of the information in the wearables domain are prerequisites for a common understanding of the information exchanged among different partners of IoT ecosystem. It would also foster semantic interoperability in the context of pan-European data exchange among public administrations, facilitate electronic interoperability among IoT businesses and also improve the statistics on IoT.

Figure 2. IoT platform vendors (adopted from Velosa et al., 2016)

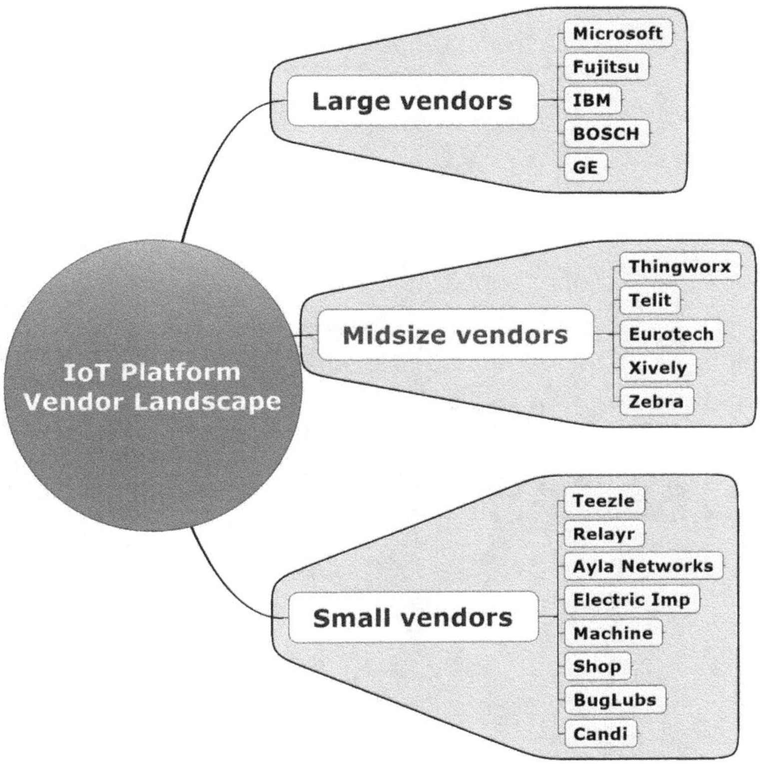

In our research, we argue that, rather than searching for a single acceptable definition, a better approach would be to develop a classification system or taxonomy (Poli et al. 2010). Taxonomies and other types of controlled vocabularies are the preferred means to achieve such a common understanding by specifying the terms of the domain, disambiguating them from each other, controlling synonyms and structuring the domain via term relationships (Yablonsky 2014, 2016a, b). For a conceptual grounding of the categories in taxonomy, we use definitions from various information resources. There are many approaches to the classification of the main entities. The current IoT standards activities include several drafts (IEEE-SA 2015; IDC 2015; Taxonomy 2016; 451 Research 2015; MacGillivray et al. 2015). However, access to this data is often inhibited by the lack of adoption of standards by the industry.

The conceptual research presented in this paper aims to collect and analyze quality data regarding the current status and prospective evolution of the IoT

Figure 3. The IOT taxonomy (technology stack)

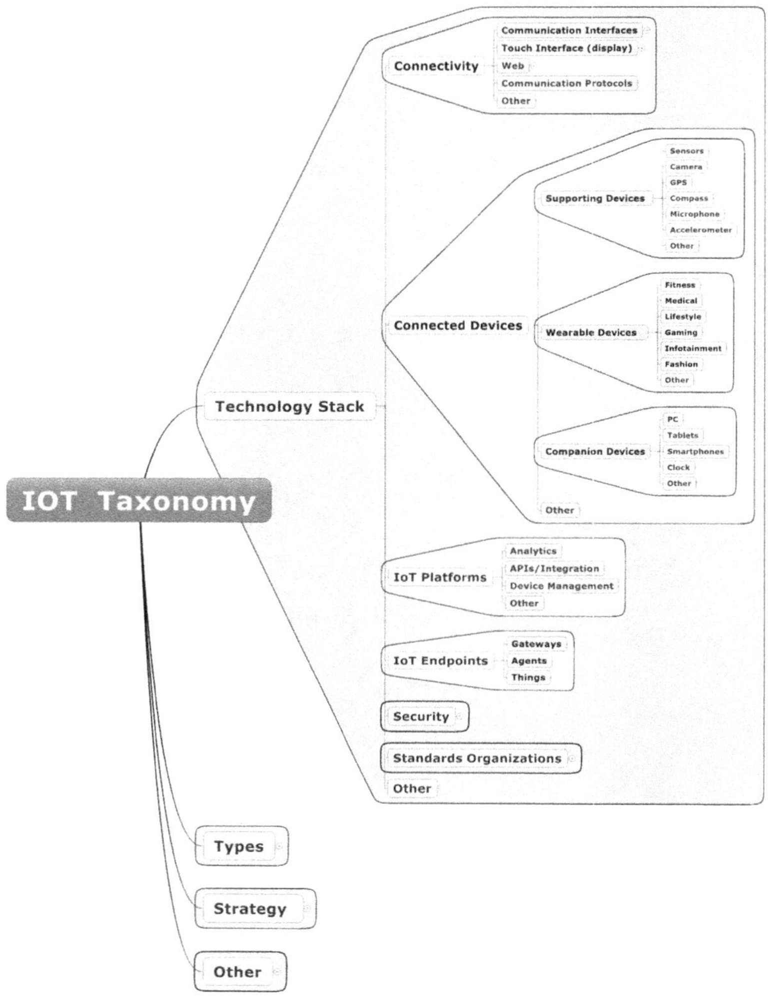

domain. The analysis proposes classification and examination of the current status of wearable technology and services both in industry and academia. The main output is the development of the set of taxonomies in the IoT domain.

A few points should be noted concerning methodology. It includes the method in which knowledge was extracted from a particular domain, the classification and organization of domain concepts, validation issues, and development tools. An appropriate knowledge source consists of academic documents that constitute the entire knowledge domain (e.g., Ugur 2013; Chana et al. 2012; Díaz et al. 2016; Fleisch 2010; Fleisch et al. 2014; Chan

Figure 4. The IOT taxonomy (types)

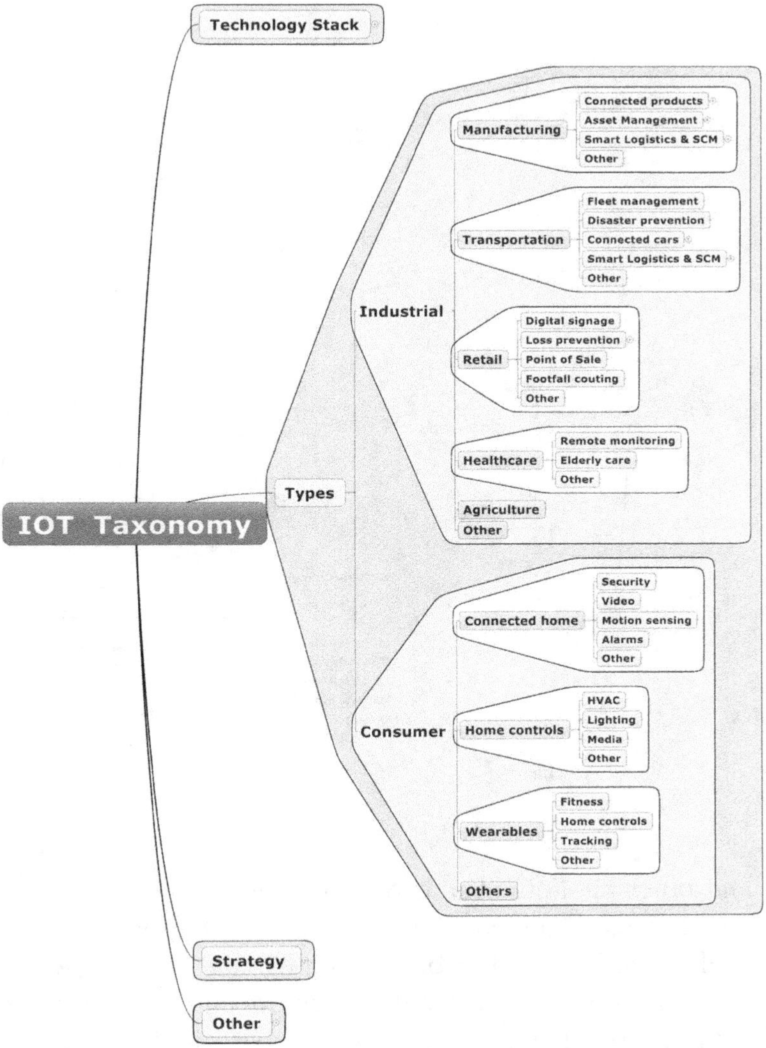

2015; Experton 2016; Green et al. 2014; Hui 2014; Hussain et al. 2014; Lee and Lee 2015; Puustjärvi and Puustjärvi 2015; Miorandia et al. 2015; Porter and Heppelmann 2014; Stock and Seliger 2016), a set of industry documents (IBM 2015; Microsoft 2015; Google 2015; Cisco 2015; Bosch 2015; Industrie 2013; IEEE-SA 2015; IDC 2015; Taxonomy 2016, Gartner 2015, 2016a, b, c), reported and observed trends and IoT activity from 2010 to 2016, IoT and wearable technology briefings, press releases, market and company's

Figure 5. The IOT taxonomy (strategy)

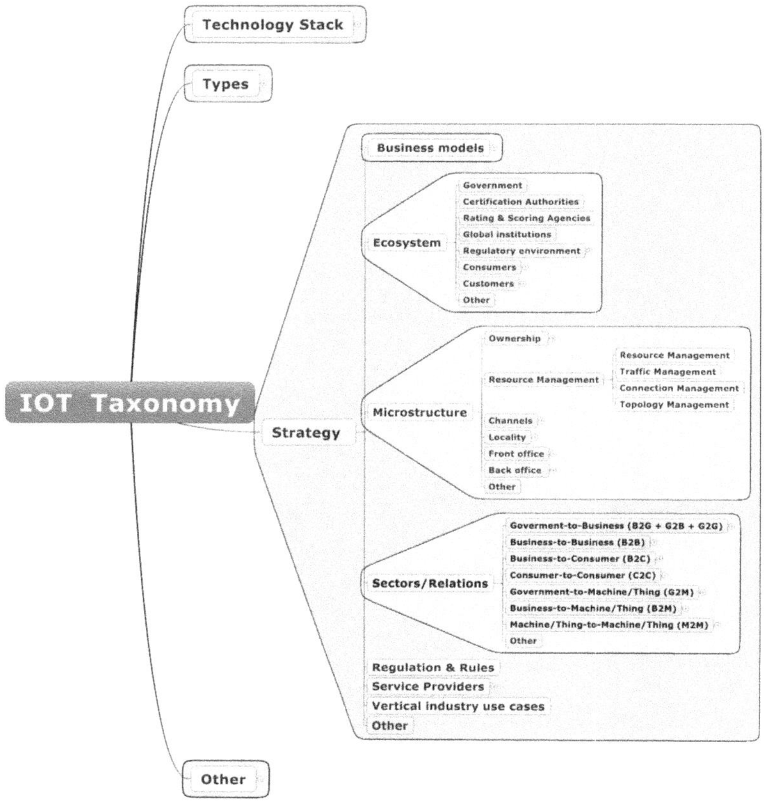

reports, and other publicly available information; web documents (blogs, web conferences, etc.) and market and company's reports.

On the other hand, depending on the nature of the knowledge domain and the taxonomy's purpose, case analysis of source documents and the validation of the first pilot version of the taxonomy by experts in the field can result in superior identification and organization of concepts.

The taxonomy is intended to provide a framework to categorize and relate technology and industry-specific aspects of the IoT research area and market. The pilot version of the taxonomy contains more than 300 terms (the hierarchy of some taxonomy concepts is shown on Figures 3, 4 and 5; a set of examples – international and local IoT and fashion companies, services, etc.); the total number of all taxonomy features is 400. So the IOT Taxonomy is too complex to be represented here in its entirety, but examples of the taxonomy parts (level 1, 2, 3) are provided in order to demonstrate both the process of classification and the result.

IoT Business Models

IoT raises a new set of strategic choices related to how the value is created and captured, how the prodigious amount of new (and sensitive) data they generate is utilized and managed, how the relationships through new digital channels with traditional business partners are redefined, and what role the companies should play as the industry boundaries are expanded (Porter and Heppelmann 2014; Hui 2014). IT and connectivity are becoming an integral part of the wearable product itself (Porter and Heppelmann 2014).

We propose this definition of an IoT MSP Platform:

A IoT Multi-Sided Platform is an organization that creates value primarily by enabling direct interactions between six (or more) distinct types of affiliated actors: IoT Producers - Software providers, IoT Producers - Service providers, IoT Producers - Advertisers, IoT Producers - Device manufactures, IoT Producers - Network providers and IoT consumers (End users, Smart Things and Businesses).

A general building blocks structure of the IoT six-sided Platform is shown in Figure 5.

Following (Osterwalder 2004; Osterwalder and Pigneur 2010; Gassmann et al. 2015; Fleisch et al. 2014; Dijkmana et al. 2015; Choudary 2015), we provide an IoT Wearable-MSP Business-Model Pattern (BPM), which is a conceptualization and formalization of the elements, relationships, vocabulary and semantics of a possible IoT MSPs for the IoT domain (Figure 6). We agree with Dijkmana et al. (2015) that the IoT boosts many other "classic" business model patterns. Up to this point, we have described components and patterns that are newly possible with the IoT Wearables in the Table 1. But the main pattern for us is MSP BMP.

IoT-Wearable MSP BMP "classically" creates value by facilitating interactions between different groups/sides of the platform. Similar to the known MSPs (Hagiu 2006; Osterwalder and Pigneur 2010; Evans 2013; Hagiu and Wright 2015; Yablonsky 2014, 2016a, b), the IoT Wearable MSP BMP Value Preposition creates value in such main areas:

- Mobilizing user groups-sides (Customer Segments) and balancing interests;
- Matchmaking between Customer Segments;
- Reducing costs by channeling transaction through the platform; and

Figure 6. IoT six-sided platform building blocks

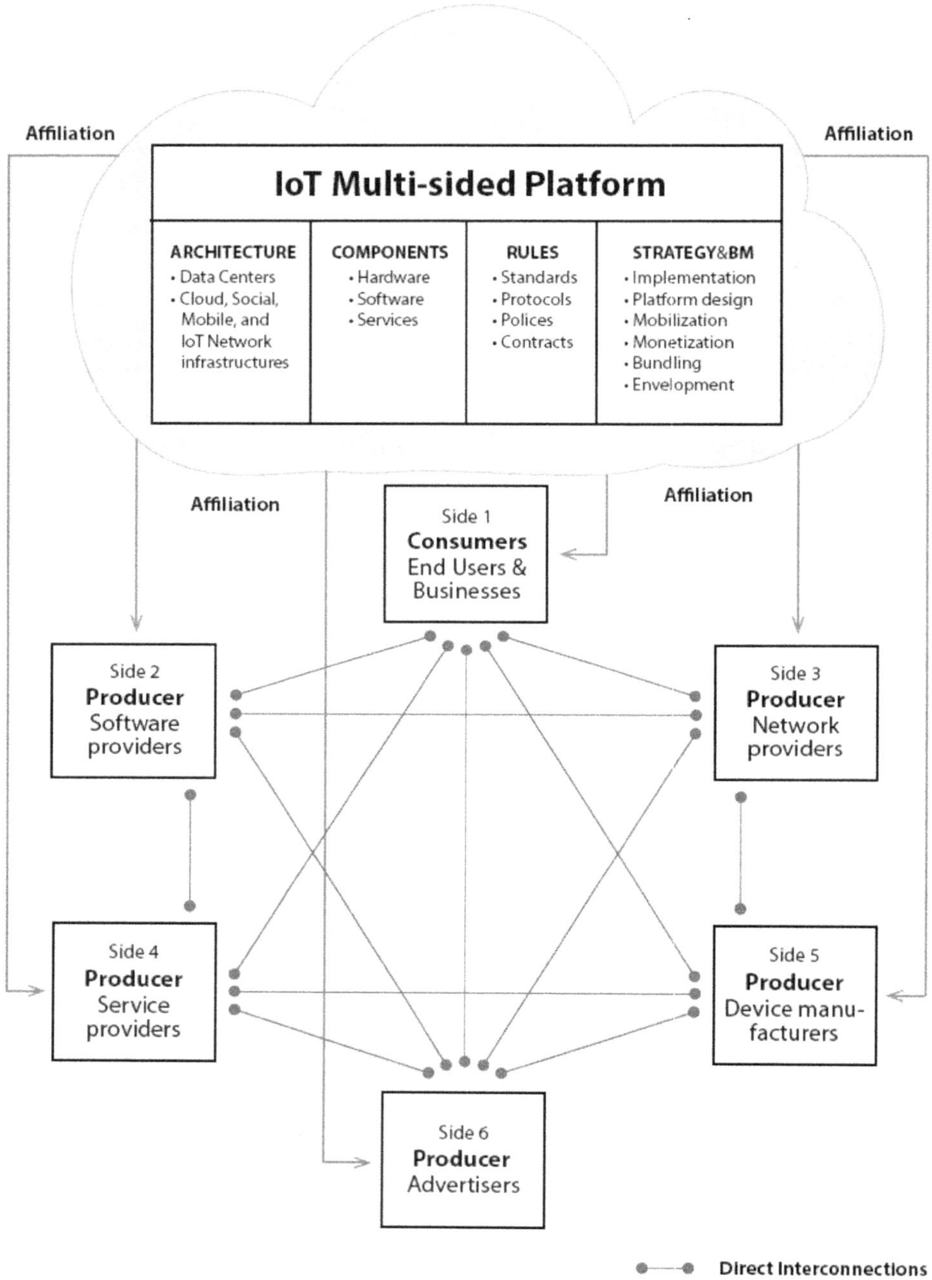

- Strategic decisions in platform evolution, multi-homing costs, scaling and liquidity.

IoT MSP BMP has three or more Platform Customer Segments, and each of them has its own Value Preposition and associated Revenue Stream. Furthermore, in some business models one Customer Segment cannot exist without another. Sometimes one or more Customer Segments may enjoy free offers or reduced prices subsidized by revenues from other Customer Segments (see Table 1). Choosing which segment to subsidize can be a crucial strategic pricing decision that determines the success of the IoT MSP BMP. IoT MSPs are of value to one group of customers only if the other groups of customers are also present.

Figure 7. IoT MSP BMP

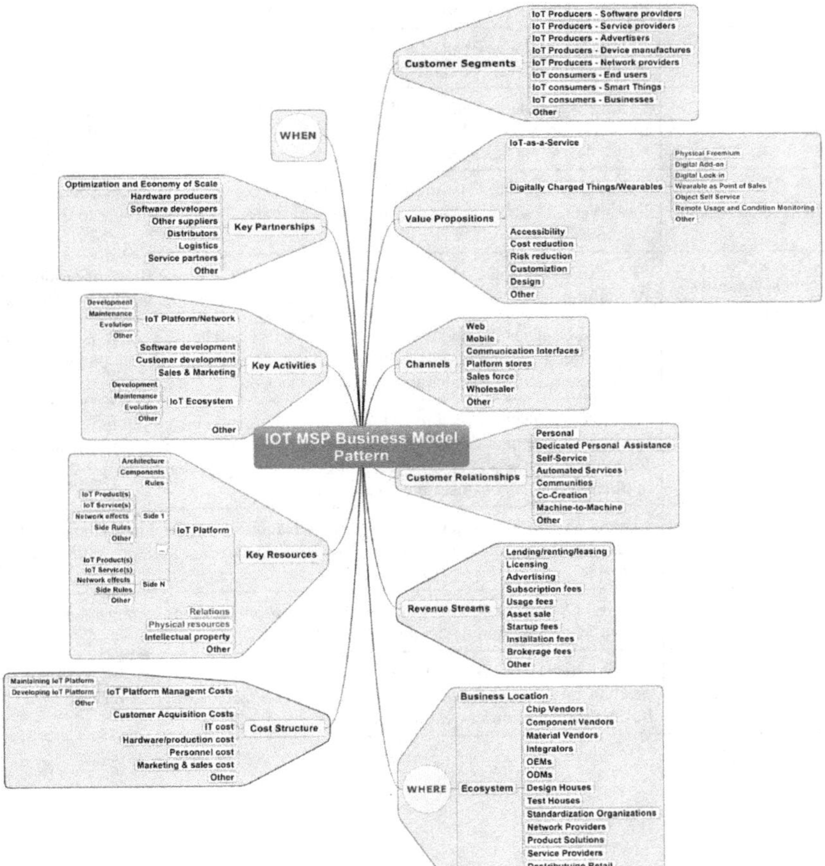

Table 1. Possible IoT BPM types (adopted from Osterwalder 2004; Osterwalder and Pigneur 2010; Gassmann et al. 2015; Fleisch et al. 2014; Dijkmana et al. 2015)

Business Model Patterns	**Facilitating Components and Patterns of the IoT Wearables**
Multi-sided Platform	"Thing/Wearable-as-a-Service" – Platforms combine IoTproducers with possible IoT consumers
Crowdsourcing	"Thing/Wearable-as-a-Service" – A "crowd" of things/wearables generates data that is monetarized. Another type of MSP
Add-on	"Digital Add-on" – Remote sale and installation of additional options for things/ wearables during the post-sale/usage period
Subscription	"Digital Add-on" – The usability of a things/wearable or sub-functions can be restricted to the time span of a subscription
Freemium	"Digital Add-on" – This business model can be applied in the physical world as well by combining free digital services with a physical thing/wearable for sale. Premium services are available for a fee
Self-service	"Thing/Wearable Self-Service" – Smart thing/wearable order consumables or services autonomously
Affiliation	"Thing/Wearable as Point of Sales" – Sales commissions for internet transactions are connected to the real world, e.g., the location of the user or a IoT wearable
Hidden revenue	"Thing/Wearable as Point of Sales" – For example, a flexible location-specific advertising becomes possible using the IoT technology
Customer loyalty	"Thing/Wearable as Point of Sales" – Customer loyalty can be rewarded not only for the purchase of a certain wearable but can be measured according to the use of the wearable, or, for instance, it's presence at a certain location
Fractionalized ownership	"Remote Usage and Condition Monitoring" – Use and consumption of wearables of lower value can be measured, making this business model applicable to those things/ wearables as well
Guaranteed availability	"Remote Usage and Condition Monitoring" – Monitoring the status of production plants or equipment via the Internet simplifies the application of the business-model pattern
Pay per use	"Remote Usage and Condition Monitoring" – Use and consumption of lower-value sensors/things/wearables can be measured too. The business model pattern is applicable to these sensors/things/wearables wearables as well
Performance-based contracting	"Remote Usage and Condition Monitoring" – Use and consumption of lower-value sensors/things/wearables wearables can be also measured. The technology for monitoring the status of production plants and equipment promotes the further application of this business model pattern
Leverage customer data	"Thing/Wearable-as-a-Service" – Things/Wearables transmit data to the manufacturer over their lifetime. The manufacturer can then use the data to improve the thing/ wearable
Lock-in	"Digital Lock-in" – Compatibility with competitors' systems is prevented by the use of a digital handshake and authentication mechanism
Razor and blade	"Digital Lock-in" – "Razor blades" can be authenticated online using wearable digital mechanisms. It eliminates the need to elaborate a safeguarding of the business model, such as through patents, for instance

The IoT MSP grows in value to the extent that it attracts more users. IoT MSP's value to a user depends on the number of other platform users and defines the Willingness-to-Pay (henceforth, WTP) for a platform participation. WTP for platform affiliation is a cap on platform fees (Eisenmann et al. 2006).

A network effect (also called network externality) is the effect that one user of a platform good or service has on the value of that product to other people (Eisenmann et al. 2006; Eisenmann 2007; Yablonsky 2014, 2016a, b). The key to IoT MSPs is to understand the network effects that occur when the value of a product depends on the number of other users (Shapiro and Varian 1999). IoT MSPs add personalization and context to the network effects between smart thing products.

The main IoT Wearable MSP BMP Costs incurred under MSP BMP relate to maintaining and developing the platform and are characterized (Yablonsky 2016a, b) by:

- Homing costs, which are related to the adoption, operation and/or other costs encountered due to platform affiliation (Armstrong 2006);
- Switching costs or the costs that consumers pay for switching from one platform to another (Shapiro and Varian 1999).

Various patterns, including IOT MSP BMP, make a free offer possible. Non-paying customers are financed by another part of the BM or by another Customer Segment of the IoT Wearable MSP BMP.

IoT MSPs enable new smart products/services/things due to the reuse of platform components. They have lower fixed costs and enable a shorter time to market for service providers.

Contemporary IoT MSP business models will have high evolutionary dynamics: the number of sides, products and/or services is steadily increasing (Yablonsky 2014, 2016a, b). A successful IoT MSP should create some network of relationships among economic entities (producers, stakeholders, dis- tributors, consumers, government agencies, etc.) that through competition and/or cooperation, facilitates the creation and distribution of a platform product or service (Yablonsky 2014, 2016a, b). The platform digital ecosystem metaphor (Weill and Woerner 2015) could be used to describe such an IoT MSP network that could be characterized by open, flexible, demand-driven, interactive, net-worked architecture and collaborative environments (Tiwana 2014; Yablonsky 2014, 2016a, b). For some IoT MSPs, the creation of a digital ecosystem is a critical success factor for positive platform dynamics.

MAIN FINDINGS AND RECOMMENDATIONS

This chapter provides the IoT taxonomy – a multidimensional framework, with a particular emphasis on technology stack, business model, product, service and platform innovations. Moreover, it presents a framework that enables analysis of IoT innovations, which is also new to the academic world, and which may serve as a starting-point for further scholar development. The goal of the IoT taxonomy is not to insist on uniformity, but rather to know and model the knowledge across the research area. In our analysis, we argue that rather than searching for a single acceptable definition of IoT domain, a better approach would be to develop a classification system of taxonomy. A clear and precise description and structuring of the information in the IoT domain are prerequisites for a common research. For a conceptual grounding of the categories in the IoT taxonomy we use definitions from different information resources. Taxonomy includes a set of basic concepts, a set of relationships hold between those concepts and a set of examples of international and local IoT platform service providers. We suggest that, although classification systems have been already used in the business and management disciplines, the more advanced quantitative methodologies have not yet been widely applied. The future research could be useful to build on these techniques in order to construct enhanced classification systems of IoT domain approaches across a variety of dimensions, in addition to the basic concepts of technology stack, business model, product, service, marketing and platform innovations.

Using the IoT MSP BMP as the basis of the IoT business models analysis, for each industry from the IoT taxonomy, we update this pattern, converting it into an IoT MSP BMP for a specific type of industry platform (healthcare IoT MSP BMP, retail IoT MSP BMP, etc.).

Finally, for a given entity – a given industry IoT MSP company – the industry MSP BMP is transferred to a company's MSP BM: *IoT MSP BMP— > industry IoT MSP BMP— > company IoT MSP BM.*

LIMITATIONS AND FURTHER DEVELOPMENTS

We think that the early stage of the IoT maturity limits our understanding of future developments in this area. From our point of view, the Standards, Protocols and Rules block in IoT MSP framework (see Figure 4) will play a key role in the future IoT-enabled innovations for wearable product/service research.

FURTHER DETAILS

Research of this chapter has been conducted with financial support from St. Petersburg State University grant (Project No. 16.23.1461.2017).

REFERENCES

Armstrong, M. (2006). Competition in two-sided markets. *The Rand Journal of Economics*, *37*(3), 668–691. doi:10.1111/j.1756-2171.2006.tb00037.x

Bosch. (2015). Retrieved December 18, 2017, from: https://www.bosch-si.com

Chan, H. (2015). Internet of things business models. *Journal of Service Science and Management*, *8*(04), 552–568. doi:10.4236/jssm.2015.84056

Chana, M., Estèvea, D., Fourniolsa, J., Escribaa, C., & Campoa, E. (2012). Smart wearable systems: Current status and future challenges. *Artificial Intelligence in Medicine*, *56*(3), 137–156. doi:10.1016/j.artmed.2012.09.003 PMID:23122689

Choudary, S. (2015). *Platform scale: How an emerging business model helps startups build large empires with minimum investment*. Platform Thinking Labs.

Cisco. (2015). Retrieved December 18, 2017, from: http://www.cisco.com/c/m/ru_ru/internet-of-everything-ioe/

Dijkmana, R., Sprenkelsa, B., Peetersa, T., & Janssen, A. (2015). Business models for the internet of things. *International Journal of Information Management*, *35*(6), 672–678. doi:10.1016/j.ijinfomgt.2015.07.008

Eisenmann, T. (2007). *Platform-mediated networks: defiNitions and core concepts*. Harvard Business School, HBS: 807–049.

Eisenmann, T., Parker, G., & Van Alstyne, M. (2006). Strategies for two-sided markets. *Harvard Business Review*, 10.

Evans, D. (2013). *Economics of vertical restraints for multi-sided platforms*. University of Chicago Institute for Law & Economics Olin, Research Paper No. 626.

Experton. (2016). *Experton industry 4.0/internet of things vendor benchmark 2016*. Retrieved December 18, 2017, from: https://www. bosch-si.com/media/en/bosch_software_innovations/documents/white_paper/manufacturing_ 3/experton_group_industry_4_0_IoT_benchmark_2016.pdf

Fleisch, E. (2010). *What is the internet of things? An economic perspective*. Auto-ID Labs White Paper WPBIZAPP-053, ETZ Zürich & University of St. Gallen. Retrieved December 18, 2017, from: http://www.im. ethz.ch/education/HS10/AUTOIDLABS-WP-BIZAPP-53.pdf

Fleisch, E., Weinberger, M., & Wortmann, F. (2014). *Business models and the internet of things*. Bosch IoT Lab Whitepaper. Retrieved from http://www.IoT-lab.ch/?page_id=10543

Gartner, J. N. (2016b.) *Technologies, architectures & vendors for the internet of things*. Gartner Webinar.

Gartner. (2014). *Gartner says the internet of things will transform the data center*. Retrieved December 18, 2017, from: http://www. gartner.com/newsroom/id/2684616

Gartner. (2015). *Research guide: the top 10 strategic technology trends for 2015*. Retrieved December 18, 2017, from: http://www. gartner.com/doc/2966917?refval=&pcp=mpe

Gartner. (2016a). *Building the digital platform: the 2016 CIO Agenda, Gartner 2016 CIO agenda survey*. Retrieved December 18, 2017, from: http://www.gartner.com/technology/cio-trends/cio-agenda/

Gassmann, O., Frankenberger, K. K., & Csik, M. M. (2015). *The business model navigator: 55 models that will revolutionise your business*. FT Press.

Google. (2015). Retrieved December 18, 2017, from: https://cloud.google.com/solutions/IoT/

Green, J., McCarson, B., & Devine, M. (2014). *Building the internet of things*. Retrieved December 18, 2017, from: https://daue6ehqissah. cloudfront.net/breakouts/2014/H-ARC-01_Cisco-Intel-IBM_FINAL.pdf

Hagiu, A. (2006). Pricing and commitment by two-sided platforms. *The Rand Journal of Economics*, *37*(3), 720–737. doi:10.1111/j.1756-2171.2006.tb00039.x

Hagiu, A., & Wright, J. (2015). Marketplace or reseller? *Management Science*, *61*(1), 184–203. doi:10.1287/mnsc.2014.2042

Hagiu, A., & Wright, J. (2015b). Multi-Sided Platforms. *International Journal of Industrial Organization*, *43*, 162–174.

Hui, G. (2014). How the internet of things changes business models. *HBR Web Article*. Retrieved December 18, 2017, from: https://hbr.org/2014/07/how-the-internet-of-things-changes-business-models

Hussain, M., Al-Haiqi, F., Zaidan, A., Zaidan, B., Kiah, V., Anuar, N., & Abdulnabi, M. (2015). The landscape of research on smartphone medical apps: Coherent taxonomy, motivations, open challenges and recommendations. *Computer Methods and Programs in Biomedicine, 122*(3), 393–408. doi:10.1016/j.cmpb.2015.08.015 PMID:26412009

IBM. (2015). *IBM Watson internet of things (IoT)*. Retrieved December 18, 2017, from: http://www.ibm.com/internet-of-things/

IDC. (2015). *IDC's worldwide internet of things taxonomy*. Retrieved December 18, 2017, from: https://www.idc.com/getdoc.jsp? containerId=256186

IEEE. (2015). *IEEE-SA internet of things (IoT) ecosystem study*. Retrieved December 18, 2017, from: http://standards.ieee.org/ innovate/IoT/

Industrie. (2013). *Recommendations for implementing the strategic initiative INDUSTRIE 4.0, Final report of the Industrie 4.0 Working Group*. Retrieved December 18, 2017, from: http://www.acatech.de/fileadmin/user_upload/Baumstruktur_nach_Website/Acatech/root/de/Material_fuer_Sonderseiten/Industrie_4.0/ Final_report/Industrie_4.0_accessible.pdf

Lee, I., & Lee, K. (2015). The Internet of Things (IoT): Applications, investments, and challenges for enterprises. *Business Horizons, 58*(4), 431–440. doi:10.1016/j.bushor.2015.03.008

LexInnova. (2015). *Wearable technology*. Retrieved from http://www.lex-innova.com/resources-reports/

MacGillivray, C., Turner, V., Kraus, B., & Kumar, M. (2015). IDC's worldwide internet of things taxonomy. Academic Press.

Microsoft. (2015). Retrieved December 18, 2017, from: https://azure.microsoft.com/en-us/services/IoT-hub/

Osterwalder, A. (2004). *The business model ontology – A proposition in a design science approach* (PhD thesis). University of Lausanne, Ecole des Hautes Etudes Commerciales HEC.

Osterwalder, A., & Pigneur, I. (2010). *Business Model Generation: A Handbook for Visionaries, Game Changers, and Challengers*. John Wiley & Sons.

Porter, M., & Heppelmann, J. (2014). How smart, connected products are transforming competition. *Harvard Business Review*, *92*(11).

Puustjärvi, J., & Puustjärvi, L. (2015). The role of smart data in smart home: Health monitoring case. *Procedia Computer Science*, *69*, 143–151. doi:10.1016/j.procs.2015.10.015

451. Research. (2015). *451 research explaining the internet of things ecosystem and taxonomy*. Retrieved December 18, 2017, from: https://451research.com/images/Marketing/IoT/IoT_Taxonomy_12.1.15.pdf

Shapiro, C., & Varian, H. (1999). *Information Rules*. Boston: Harvard Business School Press.

Stock, T., & Seliger, G. (2016). Opportunities of sustainable manufacturing in industry 4.0. *Procedia CIRP*, *40*, 536–541. doi:10.1016/j.procir.2016.01.129

Taxonomy. (2016). Retrieved from https://cantina.co/the-internet-of-things-our-take-on-a-taxonomy/

Ugur, S. (2013). *Wearing embodied emotions: a practice based design research on wearable technology*. Springer. doi:10.1007/978-88-470-5247-5

Weill, P., & Woerner, S. (2015). Thriving in an increasingly digital ecosystem. *MIT Sloan Manage Rev, 56*(4).

Y.2060. (2012). *Y.2060: overview of the internet of things, recommendation Y.2060 (06/12).* Retrieved from https://www.itu.int/rec/T-REC-Y.2060-201206-I

Yablonsky, S. (2014). E-finance innovation services in Russia. *Int J Bus Innov Res, 8*(5).

Yablonsky, S. (2016a). Multi-sided search platforms: global and local. *Int J Technol Mark, 11*(3).

Yablonsky, S. (2016b). Intermediaries in E-Commerce. In I. Lee (Ed.), *Encyclopedia of E- Commerce Development, Implementation, and Management* (Vol. 1). IGI Global. doi:10.4018/978-1-4666-9787-4.ch005

Chapter 5
Transaction Platforms:
Fintech Platforms

ABSTRACT

In this chapter the author gets involved with transaction platforms or multi-sided platforms that facilitate transactions between different types of actors. Transaction platforms are usually platform intermediaries. All platforms are digital: they capture, transmit, and monetize data over the internet and other networks, but they may as well have some physical elements included in the product offering. The author provides a platform innovation taxonomy that is a conceptualization and formalization of elements, relationships, vocabulary, and semantics of different dimensions of transaction platform innovations.

INTRODUCTION

As it was stated in Chapter 1, platforms create value in four principal ways. The first way, which corresponds to what is called *transaction platforms* or *multi-sided platforms* (MSP), facilitates transactions between different types of actors. Transaction platforms or multi-sided platforms are usually platform intermediaries. There are also *innovation platforms* that consist of technological building blocks and used as a foundation on top of which a large number of innovators or firms can develop complementary technologies, services or products (Evans and Gawer, 2016). An *integrated platform* is a technology, product or service that is both a transaction platform and an innovation platform (Evans and Gawer, 2016). *Investment platforms* consist

DOI: 10.4018/978-1-5225-5457-8.ch005

of companies that have developed a platform portfolio strategy and act as a holding company, active platform investor or both (Evans and Gawer, 2016).

BACKGROUND

According to OECD (OECD, 2010), and cited by UNESCO (MacKinnon et al., 2015) internet intermediaries bring together or facilitate transactions between third parties on the Internet. They give access to, host, transmit and index content, products and services originated by third parties on the Internet or provide Internet-based services to third parties. They offer access to a host of activities through both wired and wireless technologies. Platform intermediaries can be categorized based on function they fulfill for platform's users and industries. As it is shown (see Figure 1), platform intermediaries span a wide range of digital business activities (OECD, 2010; UNESCO, 2015).

Some examples of taxonomy platform intermediary entities are listed in Table 1.

FINTECH PLATFORMS

The financial services industry is one of the most data-rich environments on the web today; however access to this data is often inhibited by the lack of adoption of standards by the industry. The current standards activities that are likely to be of interest for the financial services industry include:

- Accredited Standards Committee X9 (ASC X9), accredited by American National Standards Institute (ANSI) that also acts as the secretariat for International Organization for Standardization Technical Committee 68 (ISO TC 68), which works on standards in the areas of payments, securities and related security technologies.
- North American Security Standards Productions Organization (NASPO), also accredited by ANSI, which works in the areas of identity proofing and verification and fraud countermeasures.
- Payment Card Information (PCI) Security Standards Council, which works on setting security standards to secure card-related data
- The Santa Fe Group about the Shared Assessments Program, which provides standards and tools for assessing IT outsource vendors

Table 1. Examples of platform intermediaries (Yablonsky, 2014a,b, 2016)

Name	Side 1	Side 2
Search MSPs		
Google	Searchers	Advertisers (other sides: developers, etc.)
Bing		
Baidu (China)		
Yandex (Russia)		
E-Bank MFPs		
Bank of America	Consumers	Merchants (other sides: government, enterprises, mobile providers, etc.)
UniCreditbank		
Citi		
Sberbank (Russia)		
e-Payment MFPs		
Paypal	Consumers	Consumers (other sides: enterprises, mobile providers, etc.)
Yandex.Money (Russia)		
E-Acquiring MFPs		
VISA	Consumers	Merchants (other sides: enterprises, etc.)
Master Card		
American Express		
E-Retail MSPs		
Amazon	Consumers (Buyers)	Enterprises (Sellers) (other sides: advertises, consumers (sellers), mobile providers, developers, etc.)
Alibaba (China)		
E-Auction MSPs		
eBay	Consumers (Buyers)	Consumers (Sellers) (other sides: advertises, enterprises, mobile providers, developers, etc.)
E-Recruitment MSPs		
Monster	Job seekers	Recruiters
LinkedIn		
E-Renting MSPs		
Airbnb	Travelers	Hosts
Game MSPs		
XBox	Consumers	Game developers
PlayStation		
E-Tourism MSPs		
TripAdvisor	Tourists	Enterprises
Social MSPs		
Facebook	Consumers	Consumers (other sides: advertises, enterprises, game providers, etc.)
Twitter		

- World Wide Web Consortium (W3C), which creates standards in web technologies, with emphasis on the emerging semantic web.

What exactly is digital finance?

Definition 1. (Yablonsky, 2014a,b).

Digital finance (or E-finance) is the provision of financial services and markets through digital channels.

Digital finance is rapidly changing the structure of financial services (Yablonsky, 2014a,b)..

Definition 2. (Yablonsky, 2014a,b)

Digital banking (e-banking, internet banking, virtual banking or online banking) refers to all forms of banking services and transactions performed through digital means. Digital banking includes systems that enable financial institution customers, individuals or businesses to access accounts, transact business or obtain information on financial products and services through a public or private network, including the Internet.

Definition 3 (Nsouli and Schaechter, 2002; Yablonsky, 2014 a,b, 2016).

Digital money (e-money or e-currency) is a stored value or a prepaid payment mechanism. The difference between digital money and digital banking is that with digital money balances are not kept in financial accounts with banks.

Definition 4 (Yablonsky, 2014 a,b, 2016).

Digital insurance (e-insurance) is the application of internet and related information technologies (IT) to the production and distribution of insurance services.

Definition 5 (Joshi, 2010; Yablonsky, 2014 a,b, 2016).

Digital trading (e-trading) is a facility that provides some or all of the following services:

- Digital order routing from users to the system.

Figure 1. Taxonomy of Transaction Platforms

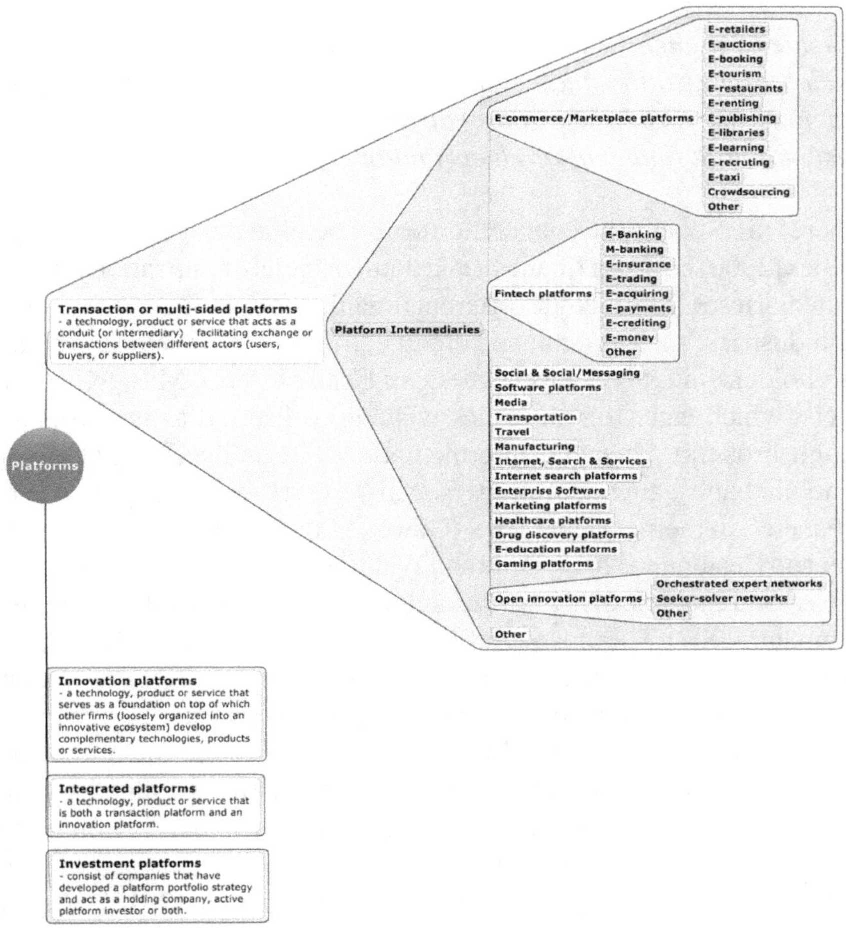

- Automated trade execution – translating orders into trade.
- Digital dissemination of bid/offer quotes and depth.
- Post-trade information (transaction price and volume data).
  Definition 6 (Yablonsky, 2014 a,b, 2016).

Digital acquiring (e-acquiring) is a service payment carried out using credit cards over the Internet.

Using the digital acquiring, companies get access to new segments of customers who prefer to pay for goods and services over the Internet using credit cards, and thus provide an additional service to their current customers.

Definition 7 (Yablonsky, 2014a,b, 2016).

Multi-sided Financial Platform (henceforth, MFP) is a financial or non-financial organization that creates value primarily by enabling direct interactions between two (or more) distinct types of affiliated customers that are using digital financial products and/or services.

There is a co-opetition (cooperation or competition) relationship (Bengtsson and Kock, 2000) between financial and non-financial organizations and digital intermediaries. MFPs include different organizations in finance, banking and other industries, such as debit and credit card payment schemes, which cater to cardholders and merchants (American Express, Visa, MasterCard), stock markets, which cater to companies wishing to list and to investors/traders (through brokers), financial intermediations, insurance, securities, online and mobile banks, auctions, shops, social services, digital payment services (PayPal etc.) in their current forms (Gawer, 2009; Hagiu and Wright, 2015b; Gawer and Cusumano, 2014; Peitz and Waldfogel, 2012; Yablonsky, 2014a,b).

A classical example of MFP is the digital acquiring (VISA) MFP Framework (Eisenmann, 2007). The most significant motivators for financial organizations to implement MFPs are the attainment of competitive advantage, tracking of technological developments and lower transaction costs.

Mobile banking (henceforth, M-banking) MFPs gives the possibility to bring basic banking and electronic transactions services to unbanked consumers in emerging and developing markets. Be that as it may in empowering multi-sided markets, m-banking solutions also provide telecommunications industry regulators with particular concerns (Anderson, 2010). Regulators need to address if the elements are in place for m-banking networks to tip towards a single platform, especially in markets with dominant operators that hold significant market share. In view of multi-homing expenses inherent in most existing m-banking MFPs, these platforms present both economic and psychological switching costs for consumers. In turn, these switching costs can have the effect of reinforcing existing network effects in markets where the incumbent already holds significant market share for voice traffic.

There are various choices accessible to telecommunication regulators in responding to the emergence of m-banking MFPs and authorities should take a measured approach to achieve optimal social and industry results. The positive effect stems from the fact that, by offering unified and easy ways to connect to common components and foundational technologies, platform leaders help to reduce the cost of entry in complementary markets and provide demand for complements, often fueled by network effects.

Playing with multisided platforms soon is going to become a fact of life for all finance and banking companies. MFPs reduce search and transaction costs and give companies vastly broader access to markets than they could achieve on their own.

Russian fintech transactional platforms will be used as a case.

Digital Banking in Russia

Russian banking sector includes 944 banks that currently operate in Russia, while dominated by state-owned giants Sberbank and VTB, as well as the regulator, Russian Central Bank. Russia's biggest banks by asset are: Sberbank, VTB, Gazprombank, Rosselkhozbank, Bank of Moscow, VTB 24, Alfabank, Unicredit Bank, Raiffeisenbank and Promsvyazbank. Foreign banks' success stories about entering Russian market include Societe Generale, UniCredit and Raiffeissenbank. Russian banking sector has a huge potential for growth, with Credit Suisse estimating that as little as 24% of Russians have bank accounts.

Alfa-Bank E-Finance Innovation Services

Global Finance announced one of the largest private banks in Russia, Alfa-Bank (http://alfabank.ru), to be the best consumer internet bank in Central and Eastern Europe in 2011. It was selected basing on the following criteria: strength of the strategy for attracting and servicing online customers, success in making clients use web offerings, growth of online customers, breadth of product offerings, evidence of tangible benefits gained from internet initiatives, together with website design and functionality.

Retail business has always been a priority for Alfa-Bank (http://alfabank.com/retail/). At the end of 2009 the total balance on individuals' accounts with Russian banks reached $247.5 billion, including $207 billion of term deposits. At that time Alfa-Bank term deposits received from individuals up to $4.037 million. If we exclude the funds obtained via the integration of Severnaya Kazna bank, Alfa-Bank's market share, in terms of cumulative balances, grew from 1.35% to 1.75%. At the beginning of 2009, Alfa-Bank ranked number seven in Russia in terms of the funds attracted from individuals. By the end of the year it had moved into the sixth place. During the reporting year, the total number of customers using Alfa-Bank's settlement and deposit services increased by 19% (up to 2,571,000). The total retail customer base across all operating segments increased by 27%, reaching 3,895,000 in 2009. This combined with a 5% reduction in staff resulted in a substantial improvement in the efficiently of the Retail Bank. The crisis sparked an increase in the popularity of credit cards among customers, increasing Alfa-Bank's credit card portfolio from $179 million to $218 million, while the number of customers using credit cards increased from 154,000 to 186,000 in 2009.

Today Alfa-Bank platform has the following main e-finance services (Figure2):

- e-Money
- e-Banking
- e-Acquiring
- e-Money
- Mobile banking ("Alpha-Mobile" including sms-banking "Alpha-Check")
- Alfa-bank cards
- e-Trading ("Alfa-Direct").

Russia's Stock Exchanges and E-Trading

Trading systems for equities, fixed income and foreign exchange are consolidating and going global. Trading is moving towards electronic platforms that are not tied to any location. Russia has two stock exchanges – the Russian Trading System (RTS) and the Moscow Interbank Currency Exchange (MICEX).

RTS (http://www.rts.ru/) runs a series of trading platforms: RTS Standard – a market for Russia's most liquid securities, traded in rubles; RTS Classica – which allows trading in both rubles and foreign currencies; RTS T+O Market – offering securities trading for retail investors with full preliminary deposit of assets and ruble settlement; RTS Board – a quote-driven market for unlisted stocks and bonds; and FORTS – a futures and options market with ruble settlement. RTS Group operates the central counterparty, the settlement securities depository and the settlement house for rubles and foreign currencies. International members of RTS include Deutsche Bank, CSFB, UBS and Morgan Stanley. Its key shareholders include global investment banks like UBS, Credit Suisse and Deutsche Bank. RTS provides collocation of internet trading systems in the data center of RTS Stock Exchange, which meets global standards (TIA-942). The collocation facility increases the reliability of services provided by trading participants to their clients and ensures the lowest possible latency for the trading system servers. Collocation results in reducing the impact of the physical environment in which data is transmitted on the speed of submitting orders and receiving market data. Twenty corporate clients accounting for roughly 40% of the total turnover on FORTS and RTS Standard markets. MICEX (http://www.micex.com/services/marketdata/profile) was founded in Russia in 1992 and trades in

Figure 2. Alfa-Bank platform services

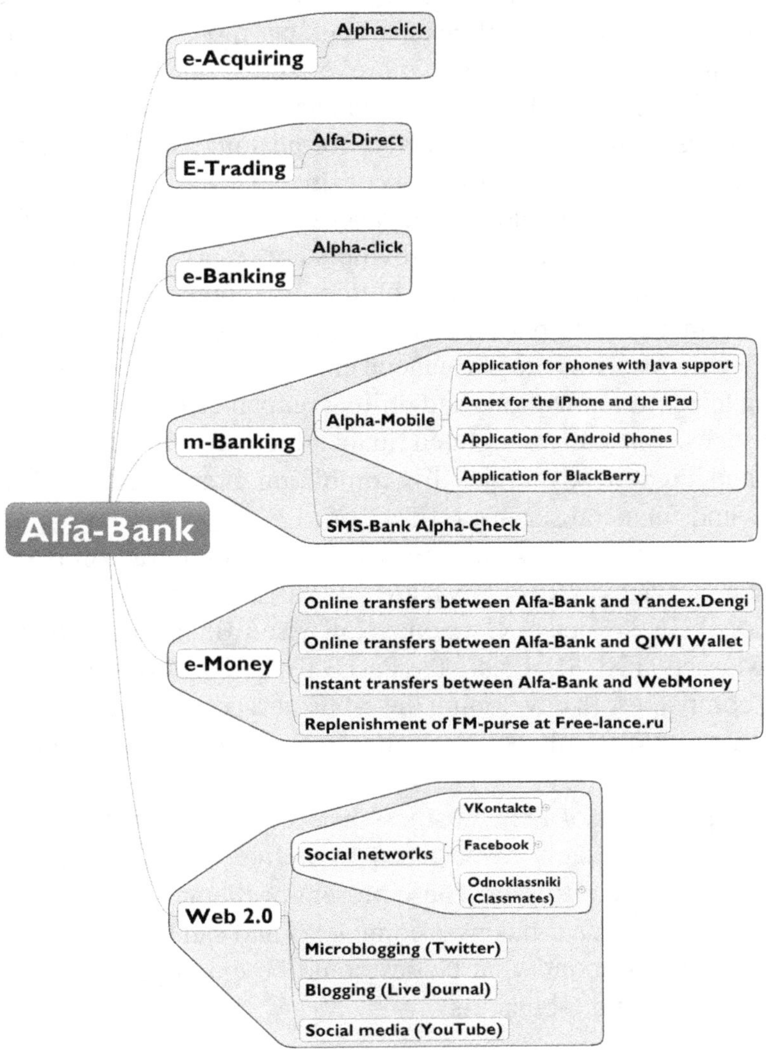

rubles. Owned by MICEX Group, it is one of the largest stock exchanges in Eastern Europe. Approximately 239 companies are listed on the MICEX, which has a market capitalization of $950 billion. As Russia's largest and most diverse exchange, every day MICEX generates new information, ranging from individual trade data, price movements to sub-index movements and corporate announcements. MICEX provides access to this data, valuable for traders, analysts, brokers, fund managers, and the media, through a range of information services (Claessens et al., 2001).

Using business data analytics techniques, such as data mining, for instance, providers can tailor products with few human input and at a very low cost. They can also better stratify their customer base and allow consumers to build preference profiles online enabling far more personalized pricing of financial services and much more effective identification of credit risks. The internet also allows new financial service providers to compete more effectively and be more effective for customers. All these forces are delivering large benefits to consumers of financial services both at retail and commercial levels. Technological advances are also changing the face of the financial services industry. New providers are emerging within and across countries, including online banks, online brokerages and companies that allow consumers to compare financial services such as mortgage loans and insurance policies. Non-profit financial entities are also entering the market, including telecommunications and utility companies that offer payment and other services. Vertically integrated financial service companies are growing rapidly and creating synergies by combining brand names, distribution networks and financial service production.

Today there are several examples of successful adoption and implementation of e-finance services in Russia by national (Yandex, Sberbank, etc.) and some international (Citi, Societe Generale, Raiffeisen Bank International, Forex Capital Markets (FXCM), FXall, PayPal) B2B, B2C and B2G e-commerce/e-banking companies. In developing the e-business, companies have faced and dealt with many difficulties present in the Russian economic and business environment.

The development of their business has required them to resolve many challenges that are likely to be faced by online commerce, retailing and banking companies in Russia. These present challenges illustrate the issues faced by many Russian e-business companies and can be used to draw some initial conclusions about what tactics could be employed to deal with the peculiarities of doing e-business.

At the moment we can outline the following features of international e-banking in Russia:

- More cross-border transactions
- Consolidation of banking and non-banking businesses
- More outsourcing of financial services.

In e-banking, outsourcing of financial services becomes easy. In other words, as customers, front-desk and back-office in e-banking could be located in different areas, a bank may outsource its back-office domestically to a low-salary state or overseas to a low-cost country.

E-Payment Services in Russia

E-money or e-currency is a stored value or a prepaid payment mechanism. The difference between e-money and e-banking is that with e-money, balances are not kept in financial accounts with banks (Yablonsky, 2014a,b). All purchases are provided by online or mobile services. From 13% to 17% of online purchases in Russia are made using electronic currencies and less than 20% are made with bank cards. According to Electronic Money, industry association founded in 2009, Russian electronic payment market, not including payments made with banking cards, reached 70 billion rubles ($2.5 billion) in 2010 and is expected to double this year.

Russian population is not unbanked, yet rather conservative about using their bank cards – about 90% of card transactions are withdrawals. There are widespread security concerns about using cards to pay online and cash-on-delivery remains the most preferred option. Electronic currency gains increasing popularity, including for buying from abroad. Consumers can fund their e-wallets, commission free, by linking a credit or debit card, topping up through ATMs or via internet banking. Cash top-up is available through a large and growing network of cash-in payment terminals throughout the country at partnering retail points, bank offices or via various money transfer services.

Russia counts plenty of payment means, however, many problems are still left unresolved:

- Users still lack trust. This is why banking cards are rarely used for online payments and Cash on Delivery (COD) is widely used in e-commerce. The lack of trust also manifests in SMS payments due to a huge number of swindling offers.

Figure 3. Yandex.Dengi taxonomy

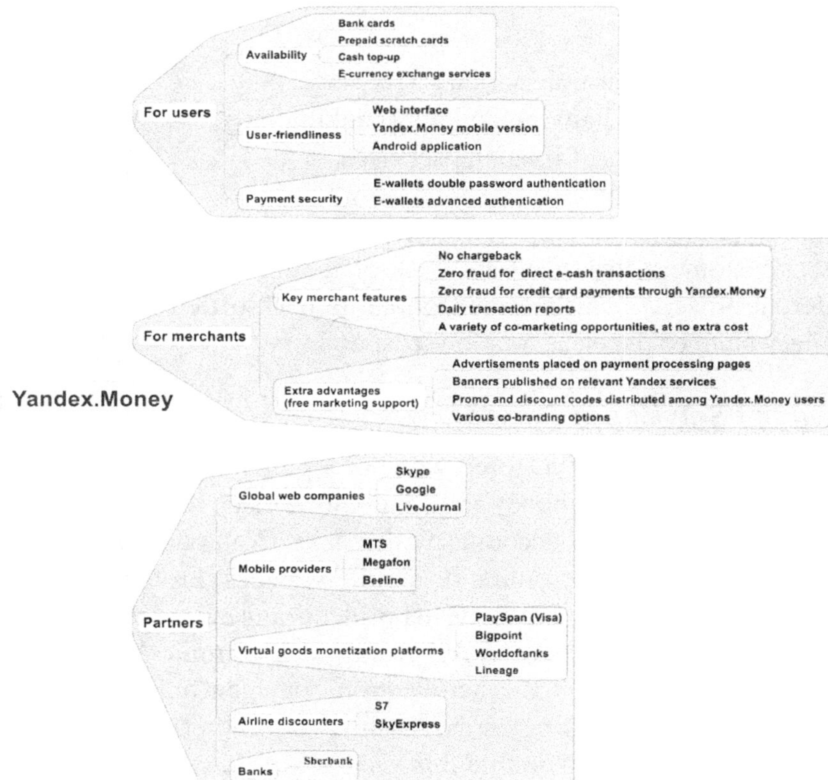

- Alternative solutions for online payments complicate user experience. Using electronic currencies requires several, sometimes uneasy steps. Offline terminals are user friendly but interrupt the online buying experience.
- Users and/or merchants are often charged with significant fees. Fees are unreasonably high with telephone-based payment means; they can be significant with offline terminals and electronic currencies as well.
- Pre-authorized, recurring payments are possible only in few cases and are still used very rarely. This complicates the business models of several industries.
- Merchants have to aggregate a great number of payment solutions, which leads to complex processes or to the necessity of paying for additional aggregation services.

The main Russian electronic currency operators are: WebMoney, Yandex.Dengi, RBK Money, MoneyMail, Cyberplat and Rapida. Yandex.Dengi (Yandex.Money) and WebMoney are the two top Russian operators. In 2010, the number of e-wallets in Russia grew by 50% (Source: Russian Electronic Money Association, March 2011). According to the latest research, 78% of the online audience have heard of Yandex.Dengi, and 61% have made one or more payments with Yandex.Dengi during the last six months (Source: TNS Media Research, Moscow audience, aged 18–45, February 2011).

The Yandex.Dengi`s e-payment system (Figure 3) was founded in 2002 to enhance commercial effectiveness in the internet industry and is a 100% subsidiary of Yandex, the leading Russian-language search engine. Yandex.Dengi is the first e-payment system in Russia that allows users to fund their e-wallets by linking a credit or debit card, at no commission. In 2011, Yandex.Dengi successfully launched service in the Commonwealth of Independent States (CIS). Users from Ukraine, Belarus, Kazakhstan and other CIS countries are now eligible to use Yandex.Dengi. The service is commission-free, but for security reasons Yandex is currently limiting the purchases to 5,000 rubles per day and 30,000 rubles per month ($172 and $1,034, respectively). For legal reasons, the service is not available for non-residents nor holders of cards issued by banks that are not registered in Russia.

The users of Yandex.Dengi can easily top up their e-wallets, and then transfer Yandex.Dengi to other users, spend them online or withdraw cash. Yandex.Dengi scratch cards are sold at all Russian Post offices in every city, town and village in Russia, which makes this top-up method utterly accessible. Prepaid scratch cards are available to buy throughout the country and can be activated online. Top-up amounts range from 300 to 10,000 rubles. Yandex.Dengi users can use cards issued by any bank operating in Russia to top up through ATMs and payment terminals of Sberbank, Russia's largest bank with about 28,000 ATMs and 17,000 payment terminals across the country. E-currency exchange services, such as WebMoney, RoboxChange, Ukash, etc., conveniently convert funds into Yandex.Dengi. It may be distinguished by an easy-to-use web interface, a universal accessibility online and a fast checkout.

By default, e-wallets are secured with a double password authentication required for every online payment. Consumers can opt for even stronger protection with advanced authentication – instead of a single payment password, they can use a table of codes or an e-token to generate unique passwords for each transaction. For Merchants Yandex.Dengi is the fastest, most secure and reliable way for e-businesses to collect payments, and an effective solution for targeting the growing markets of Russia and CIS. Yandex.Dengi offers a simple and efficient model, with advantages over credit card processing. It can help e-businesses successfully enter the Russian market, attract new customers and sell more by introducing products to thousands of users daily. In general, this model, as well as the one of WebMoney (see Figure 4), have many characteristics in common with the one that PayPal has been offering its international users for years.

Qiwi, the leading Russian operator of offline electronic terminals (ATM), offers a similar service to the users who have registered online. Their e-wallet system can also be linked to bank cards, but the service takes a commission of a 2.25% of the purchase amount. The Qiwi daily purchase limit is 150,000 rubles ($5,172).

Web 2.0 E-Payment Services

Russian social networks Vkontakte and Odnoklassniki launched their internal currency in 2011. Now users can pay for virtual and real network offered goods using social networks virtual currency. Vkontakte had created a merchant API that allows users to click on the button to have the goods they like on the merchant sites added to their wish-lists on their profiles inside the network (http://valuedrivenbanking.blogspot.com/). And later these items could be purchased through a special sequence with money deducted either from a credit card specified by the user, or from a virtual currency account. For that matter, Vkontakte offered a special top-up scheme with QIWI cash kiosks. Vkontakte network has recently passed through a strict PCI DSS certification process. Now it can be used to store card payments data that allow the customers to pay through banking cards. Almost the same has been offered by Odnoklassniki where the cards' linking is supported by Alfa-Bank.

Mobile Payment Services

Mobile payment services (Figure 5) allow operators, banks and financial institutions to introduce new service offerings such as micropayments, bill pay and top-ups. This payment method makes it possible to purchase goods from merchants, top-off credit balances for airtime or make person-to-person payments, loan payments, bill payments and more. Thus, it introduces a new channel to interact with customers. Usually m-payment services are adapted to all networks, mobile devices, languages, currencies and deliver communications across multiple mobile channels, including SMS, USSD,

Figure 4. WebMoney taxonomy

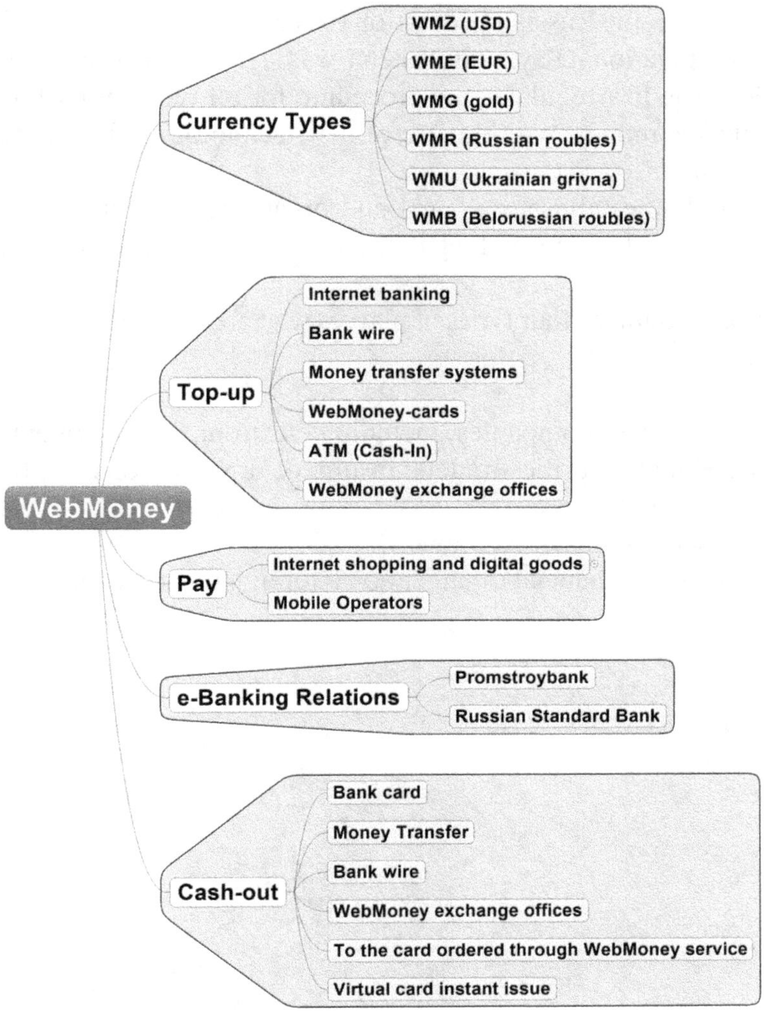

mobile web browser and mobile client applications. Financial institutions in developed and emerging markets can rely on m-payment services for a complete mobile banking solution that supports all mobile transactions conducted by a mobile phone. In Russia WebMoney Transfer and Yandex.Dengi have mobile versions. iPhone and Android applications are also available. Consumers all over the country can top up their Yandex.Dengi e-wallets, commission-free, at any of Euroset chain stores, the largest mobile phone retailer in Russia.

Government Regulations

In summer 2011, the Russian Federation Federal Law No. 161-FZ of June 27, 2011, 'On the National Payment System' was signed in by Russian President Medvedev. The law regulates the procedure for all types of cash payments with electronic money, with the use of new technologies, including mobile communications.

The new law establishes a legal and organizational foundation for the country's payment system and regulates the procedure for rendering payment services.

The law recognizes four types of players who are entitled to offer payment services:

- Money transfer companies, including electronic money operators;
- Professional securities traders, insurance organizations and the bodies of the Federal Treasury;
- The federal mail service;
- International financial organizations, foreign central banks and foreign banks.

Figure 5. Mobile payments taxonomy

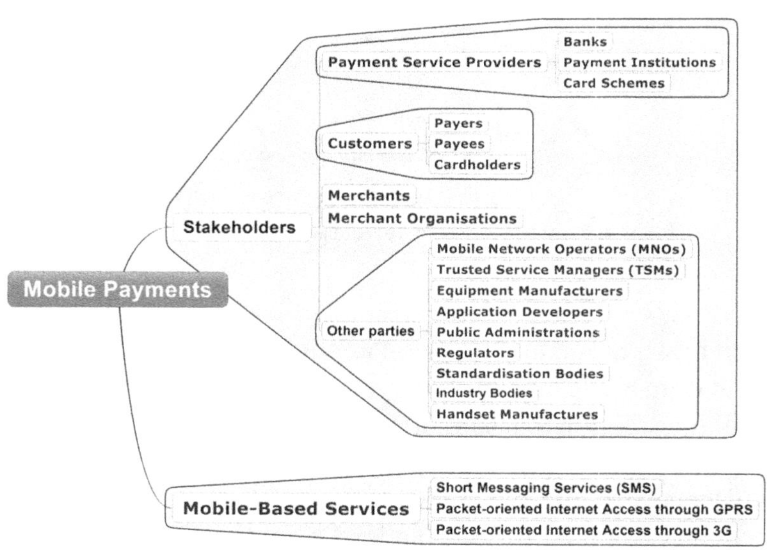

The law concerns, in particular, payments made with bank cards – Russian bank cards market estimated to be worth about $90 billion in 2009 – and electronic currencies like Yandex.Dengi and WebMoney, as well as systems for making payments through mobile devices.

The new rules also affect payments via electronic terminals, these ubiquitous offline machines and virtual wallets through which Russians pay virtually for everything from mobile phone bills to purchases made at e-commerce websites, utilities, taxes, and fines. This thriving market is currently dominated by Qiwi, followed by Cyberplat, Euroset and Eleksnet. The law stipulates that electronic payment activity can be conducted by credit organizations, including non-banking ones, which meet minimum capital requirements set at 18 million rubles, approximately $640,000. No specific license for operating payment services is required, but electronic payment operators have to inform the Russian Central Bank about their activity.

In early discussions, Russian lawmakers had considered prohibiting transmission of payment-related information to other countries, greatly curtailing international payment transactions. This move would have forced such companies as Visa and MasterCard to have processing centers located on the Russian territory to facilitate international transactions. This proposition, however, was opposed by the Ministry of Finance and a compromise was found. The new legislation allows payment processing to be organized in any country in line with service continuity guarantees, but requires operations and settlement centers to be located on the Russian territory.

FUTURE TRENDS

The key drivers of the evolution of e-intermediaries are based on the following building blocks:

- **Technology:** Such ICT based innovations (Gartner, 2015) are prime enablers for new e-intermediaries' opportunities:
    - Computing everywhere will change how people experience the world
    - The Internet of Things brings the power of device ecosystems to e-intermediaries' MSPs.
    - Advanced, pervasive and invisible analytics will enable every platform to deliver actionable insights to consumers.

- 3D printing is approaching a critical inflection point in new digitization of platform products and services
- Cloud computing will emerge to unify cloud and mobile platform strategies.
- Software-defined architecture for platform infrastructure and services is required for dynamic, agile, flexible systems to support digital platforms.

- **Globalization:** Worldwide liberalization of trade and investment can enable the growth of global business including e-business.
- **Regulations:** Proper regulations of the global digital economy can facilitate the growth in key areas. For example, European Commission sets out building blocks for a Digital Single Market (Digital Single Market, 2015). The Commission's Digital Single Market Strategy for Europe is based on three pillars:
  - Better access for consumers and businesses to online goods and services across Europe;
  - Creating the right conditions and a level playing field for digital networks and innovative services; and
  - Maximizing the growth potential of the digital economy.
- **Competition and Globalization:** The above factors created a globally fertile and competitive environment and pool of talents to compete for introducing new technologies, concepts and models.
- **Capital:** State and private capital can provide the financial means to put these technical and regulation wheels in motion.

CONCLUSION

Digital intermediaries enable the Internet to drive economic, social and political development, for instance, by facilitating aggregation of demand, new models of collaboration, citizen journalism and civic participation (OECD, 2010; MacKinnon et al., 2015).

E-intermediaries form the following main groups:

- Internet access and service providers (ISPs).
- Data processing and web hosting providers, including domain name registrars.
- Platform intermediaries.

Currently, the dominant trend of intermediary business models refers to platform e-intermediaries MSPs that form two-sided and multi-sided markets.

Intermediaries MSPs are described in details in the proposed taxonomy. The role of intermediary MSPs is to:

- Minimize transaction costs through matchmaking and audience making;
- Minimize costs through the elimination of duplication;
- Permit value-creating exchanges that would not take place otherwise;
- Enforce innovation and competition.

The study of intermediary MSPs is important because it is a large and growing share of global economy and MSP companies face distinctive management challenges: familiar rules such as value-based pricing may fail and traditional barriers to entry may no longer hold.

REFERENCES

Anderson, J. (2010). M-banking in developing markets: Competitive and regulatory implications. *Info*, *12*(1), 18–25. doi:10.1108/14636691011015358

Digital Single Market. (2015). *European Commission Digital Single Market. Bringing down barriers to unlock online opportunities*. Retrieved July 20, 2015, from http://ec.europa.eu/priorities/digital-single-market/

Eisenmann, T. (2007). *Platform-Mediated Networks: Definitions and Core Concepts*. Harvard Business School, HBS 807-049.

Evans, P. C., & Gawer, A. (2016). *The Rise of the Platform Enterprise A Global Survey*. New York: Center for Global Enterprise. Retrieved December 18, 2017, from: http://www.thecge.net/wp-content/uploads/2016/01/PDF-WEBPlatform-Survey_01_12.pdf

Gartner. (2015). *Research Guide: The Top 10 Strategic Technology Trends for 2015*. Retrieved July 20, 2015, from http://www.gartner.com/doc/2966917?refval=&pcp=mpe

Gawer, A. (2009). *Platforms, markets and innovation*. Edward Elgar Publishing. doi:10.4337/9781849803311

Gawer, A., & Cusumano, M. A. (2014). Industry platforms and ecosystem innovation. *Journal of Product Innovation Management*, *31*(3), 417–433. doi:10.1111/jpim.12105

Hagiu, A., & Wright, J. (2015b). Multi-Sided Platforms. *International Journal of Industrial Organization*, *43*, 162–174. doi:10.1016/j.ijindorg.2015.03.003

MacKinnon, R., Hickok, E., Bar, A., & Lim, H. (2015). *Fostering Freedom Online. The Role of Internet Intermediaries, A report prepared for UNESCO's Division for Freedom of Expression and Media Development.* Retrieved July 20, 2015, from http://unesdoc.unesco.org/images/0023/002311/231162e.pdf

OECD. (2010). *The Economic and Social Role of Internet Intermediaries.* OECD Digital Economy Paper No. 171.

Peitz, M., & Waldfogel, J. (2012). *The Oxford Handbook of the Digital Economy.* Oxford University Press. doi:10.1093/oxfordhb/9780195397840.001.0001

Yablonsky, S. (2014a). E-finance innovation services in Russia. *International Journal of Entrepreneurship and Innovation Management*, *8*(5), 523–551.

Yablonsky, S. (2014b). ICT-based Financial Innovations. In A.-L. Mention & M. Torkkeli (Eds.), *Innovation in financial services: a dual ambiguity* (pp. 184–224). Cambridge Scholars Publishing.

Yablonsky, S. (2016). Intermediaries in E-Commerce. In *Encyclopedia of E-Commerce Development, Implementation, and Management* (pp. 48–73). IGI Global. doi:10.4018/978-1-4666-9787-4.ch005

KEY TERMS AND DEFINITIONS

Digital or Electronic Business (E-Business): E-business usually refers to a business conducted over the internet, but in general, any network, public or private, can be included. E-business is a term that includes servicing customers, collaborating with business partners, and conducting electronic transactions within an organization, as well as buying and selling.

E-Acquiring: The service payment carried out using credit cards over the internet.

E-Insurance: The application of internet and related information technologies to the production and distribution of insurance services. In a narrower sense, it can be defined as the provision of an insurance cover whereby an insurance policy is solicited, offered, negotiated, and contracted online.

Electronic Banking (E-Banking, Internet Banking, Virtual Banking or Online Banking): All forms of banking services and transactions performed through electronic means. E-banking includes systems that enable financial institution customers, individuals or businesses, to access accounts, transact business, or obtain information on financial products and services through a public or private network, including the internet.

Electronic Finance (E-Finance): E-finance is a part of e-business and provides financial services over the internet, but in general, any network, public or private, can be included.

E-Money: The monetary value that is stored on electronic devices and which is acceptable as a means of payment by undertakings other than the issuer's.

E-Trading: The facility that provides some or all of the following services: electronic order routing from users to the system; automated trade execution—translating orders into trade; electronic dissemination of bid/offer quotes and depth; post-trade information (transaction price and volume data).

Internet Intermediaries: Internet intermediaries or e-intermediaries can be defined as organizations bringing together or facilitating transactions between third parties on the internet.

Multi-Sided Financial Platform (Henceforth, MFP): Multi-sided financial platform is a financial or not financial organization that creates value primarily by enabling direct interactions between two (or more) distinct types of affiliated customers that are using e-financial products and/or services.

Multi-Sided Platform (Henceforth, MSP): Multi-sided platform is an organization that creates value primarily by enabling direct interactions between two (or more) distinct types of affiliated customers. Payment systems are classic examples of shared cost-reducing MSPs – they provide an infrastructure that significantly eases transactions between buyers and sellers by eliminating the need for barter.

Taxonomy: Taxonomy is comprised of a hierarchy of concepts linked by a transitive subsumption relation (often called isA or subClassOf) whereby each instance of a class can be inferred to be an instance of all parent classes. Taxonomies are strict hierarchies: each class has at most one parent.

Chapter 6
Integrated and Transactional Platforms:
Search Platforms

ABSTRACT

This chapter aims to collect, analyze, and compare quality data regarding the current status and prospective evolution of search engine platforms offered by leading platform companies. The authors describe the search MSPs competition across different markets in a general way and compare several emerging markets where Google MSP is not a leader. The methodology and organization of search platform ontology/taxonomy and search platform business models are discussed. Then detailed comparative analysis of search MSP innovation services and strategies is developed for Google SMSP and Yandex SMSP. Finally, the possible ways of further evolution of local SMSP and services are discussed.

INTRODUCTION

The qualitative research presented in this chapter aims to collect and analyze quality data regarding the current status and prospective evolution of search MSP. The analysis proposes classification and examination of the current status of global and local multi-sided search market, and possible ways of further evolution of search innovation technology and services.

DOI: 10.4018/978-1-5225-5457-8.ch006

Search platforms could be classified in one of the two platform type groups: integrated and transactional platforms (see Figure 1).

Definition of the search MSP (Yablonsky, 2016):

Search Multi-sided platform (henceforth, SMSP) is an organization that creates value primarily by enabling direct interactions between two main (or more) distinct types of affiliated customers: web/mobile customers and advertisers.

Today SMSP usually has more than two distinct types of affiliated customers, including web/mobile surfers, advertisers, developers, enterprises etc., that form SMSP ecosystem. Today, the SMSP competition is on a much larger scale. It is a competition between vast SMSP ecosystems made up of hardware, software and online services.

The typical role of SMSP intermediaries is to:

- Minimize transaction costs through matchmaking and audience making.
- Minimize costs through the elimination of duplication.
- Permit value-creating exchanges that would not take place otherwise.
- Enforce innovation.

As well known, Google is the leading search MSP in the world, with a global share of search traffic around 70,8%. Google is the leader in search in every European country except for the Czech Republic and Russia. Traditionally the leaders in Japan and China have been Yahoo! and Baidu, respectively. Another large and growing Asian market, India, is however largely dominated by Google.

As on any market, competition in the search is good and generally results in stronger products that benefit consumers. All search engines provide multiple results to any given query, but different search engines generally produce a different set of results and a different user experience. Search MSP customize their results to web users with different geographic locations and languages, but the extent, nature and impact of these differences on MSP's success are unknown. However, the theoretical debate on the role of market leaders in multi-sided search markets is still limited: most of the literature on multi-sided search markets is focused on monopolistic pricing and symmetric

Figure 1. Integrated and transactional platforms

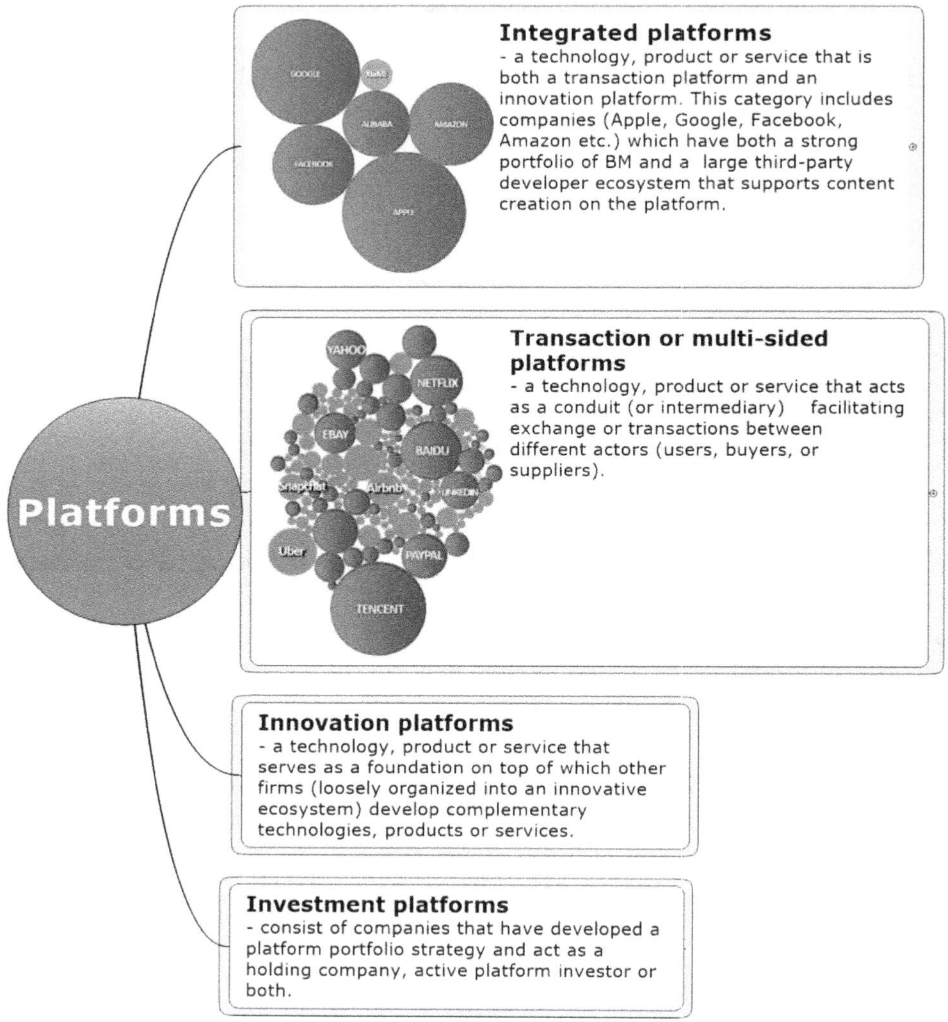

competition between platforms, not on competition between a potentially dominant platform and its followers (Etro, 2013).

SEARCH PLATFORM MARKET

In particular, a wide interest has been focused on the market for search and display advertising that is created by search platforms. In 2014 this market was worth more than $50 billion in advertising. In most local markets, users have little real choice in search engines and search platforms. According to ComScore, Google and its affiliated websites comprise 67.6% of the search engine market share in the US, and, according to Netmarketshare – 66.44% worldwide. Though noticeable, Google is not the only search engine available. There are numerous other engines that provide various interfaces, search algorithms and other unique features. All search engine platforms provide multiple results to any given query, but different search engines generally produce a different set of results and a different user experience. In descending order, the remaining most popular search engine companies in the US, by market share, after Google, are Microsoft (18.7%), Yahoo (10.0%), Ask Network (2.4%), and AOL (1.3%), according to ComScore report.

There are mostly six big-scale SMSPs in the world, and very few countries where the landscape isn't dominated by a single search player – Google, with a more than 90% market share:

- **China:** According to Chinainternetwatch.com, the biggest player in Chinese search engine market is not Google but Baidu, with about 70% market share, followed by Qihoo 360 and Sogou/Soso league.
- **Japan:** According to Semjapanese.com, in 2013 Yahoo! Japan (independent entity from Yahoo! Corporate) led the search engine market with a 56% share, Google – 40%, others – 4%.
- **Hong Kong:** Yahoo!, exact market share unavailable.
- **Russia:** Yandex, the leading Russian search engine, is the fourth-largest one in the world (Yablonsky, 2012). Reports by Bloomberg and Yandex show that in the first-quarter of 2014 Yandex's share averaged 61,9%, with Google's share around 26% and Mail.ru – the other major search platform in Russia, at about 3,1%, and others – at about 9%.
- **Czech Republic:** Seznam, 45%.
- **South Korea:** Naver, 73%.

The structure of the market for search advertising depends on a particular form of network externality which is quite different from the one emerging in other markets (Etro, 2013). This is due to the two main reasons: one related to technological issues and the other – to the role of multi-homing by advertisers.

Figure 2. The search MSP building blocks

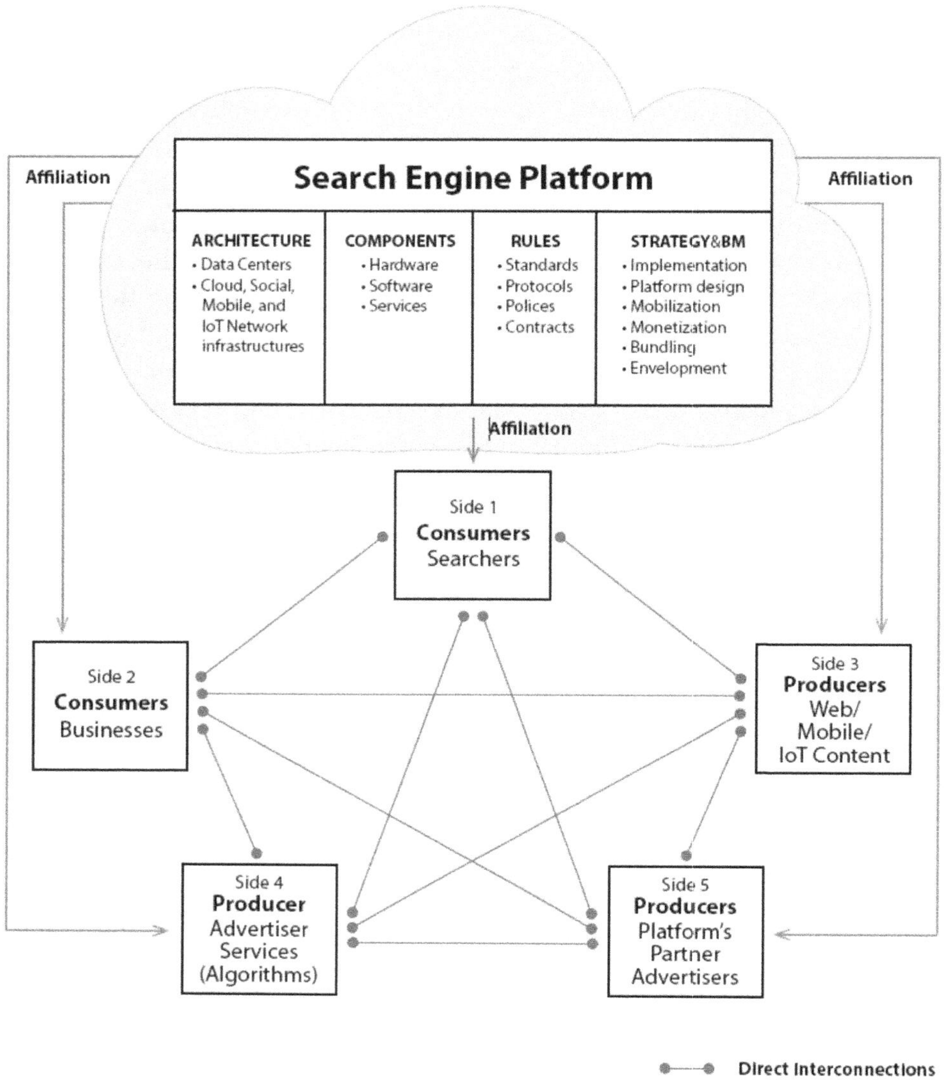

SEARCH MULTI-SIDED PLATFORMS

Usually, to win in certain markets, local knowledge is the key.

The search MSP building blocks are shown on the Figure 2. As it can be seen, a SMSP has minimum 3 sides: content providers (who need users), users/searchers (who search for content) and advertisers (who want access to users).

Figure 3. Search MSP BMP (2 levels)

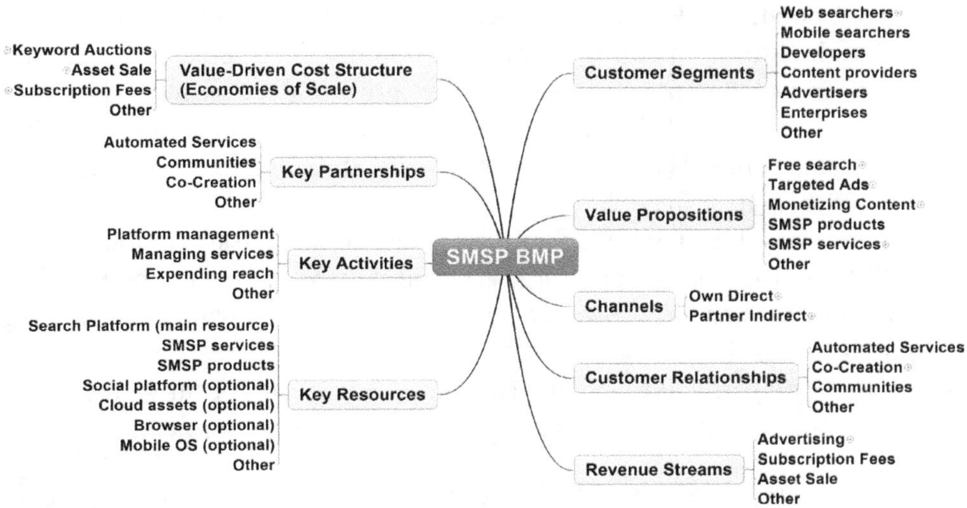

For a clear and precise description and structuring of the information in the SMSP domain we propose the SMSP Business Model Pattern, as shown on the Figure 3.

Google and Yandex SMSP Comparison

In the present work we analyze only Russian emerging search markets where Google MSP is not the leader. According to the numbers released by TNS Russia and comScore (2010), the top websites in Russian market are:

- Yandex with 26 million monthly visits.
- Mail.ru and Rambler, which dominated the market in the late 1990s, with 25 million monthly visits. Mail.ru Group, founded in 1998, is the leading Internet portal that offers a webmail service, a social network of its own (MoiMir.ru), a popular search engine and many online games. Mail.ru Group recently filed for a $876 million IPO in London Stock Exchange.
- Vkontakte.ru (vk.ru) with 20 million monthly visits. Vkontakte.ru was founded in 2004 and is the most popular Russian social network. It can also be described as an almost direct copy of Facebook. Its reach has grown most dramatically in the past two years from 30% to 73% of Internet users. Interesting to note that the 3 Russian social networks

(Vkontakte.ru, Odnoklassniki.ru and Moi-Mir.ru) are almost ten times more popular in Russia than Facebook, according to comScore.

- Google Sites with almost 20 million monthly visits. Google is the only foreign web company whose user base grew in the past two years from 41% to 67% of all Russian Internet (henceforth, Runet) users. Other companies like AOL LCC and Microsoft sites decreased their reach by more than 10% since 2008.
- Odnoklassniki.ru with 16 million users. Odnoklassniki, founded in 2006, is Russia's second-largest social network after Vkontakte. The website's reach has grown by a mere 8% in the past two years, reaching a 38% of all Runet users.
- Other players like Nigma.ru and Quitura.ru occupy relatively insignificant positions, in spite of their innovative approach.

Yandex and Google SMSPs have based their business models on advertising revenues. The key factors behind Yandex's early success in Russia was understanding of Russian culture and peculiarities of the Russian language in the search engine, as well as the abundance of successful web services for Russia, including:

- **Vertical Search:** Images, video, music, maps with traffic jams, news, people search, product search (Yandex.Market) etc.;
- **Services:** Mail, photos, calendar, payment system (Yandex.Money), web-analytics system (Yandex.Metrika) and many others.

Yandex MSPS's key resource is its search platform, whih powers three different services:

- Web search (Yandex.com),
- Advertising (Yandex.Direct), and third-party content monetization (Yandex ad network).

These services are based on highly complex proprietary search and matchmaking algorithms supported by an extensive IT infrastructure. Organic and paid Search Results in Yandex contain both organic results and paid contextual links, vertical search information, etc. By the way, Yandex is the only search engine that offers users to compare the search results from other search engines directly on the search results page. This way the developers probably evaluate the quality of search results.

The heart of Yandex SMSP's business model is the value proposition of providing extremely targeted online text advertising in Russia, Ukraine, Belorussia, Kazakhstan and Turkey. Through service called Yandex.Direct advertisers can publish advertisements and sponsored links on Yandex's search pages (and on affiliated content network, as we will see later). Yandex. Direct is an auction-based advertising placement service, which uses the most advanced auction theory and relies on distributed infrastructure to process millions of auctions every day. Yandex.Direct lets advertisers deliver relevant text-based ads, targeted at particular search queries or content on Yandex websites or third-party websites in the Yandex ad network, in a cost-effective way. Yandex.Direct enables advertisers to present ads to users at the precise moment they are looking for information related to the advertiser's product or service.

When people use Yandex search engine, the adds are displayed alongside the search results. Yandex ensures that only the ads relevant to the search term are displayed. The service is attractive to advertisers because it allows them to tailor online campaigns to specific searches and particular demographic targets. The model only works, if many people use Yandex's search Engine. The more people Yandex reaches, the more ads it can display and the greater the value created for advertisers.

Yandex SMSP has a very distinct revenue model. It makes money from only one Customer Segment, the Advertisers (first side of SMSP), while subsidizing free offers to the two remaining segments: Web surfers (second side of MSP) and Content owners (third side of SMSP). This is logical because the more ads it displays to Web surfers, the more it earns from advertisers. The substantial revenue that Yandex earns from Yandex.Direct allows it to continuously improve its free offers to search engine and Yandex ad network partners users. As Yandex's value proposition to advertisers depends heavily on the number of customers it attracts to its website, Yandex caters to this second group of consumer customers with powerful search engine and a growing number of tools such as Yandex.Mail (Web based e-mail), Yandex.Maps, and Yandex.Fotki (an online photo album), as well as Yandex specialized and vertical search engines, classified according to their relative relevance, such as Yandex.News, Yandex.Market, Yandex.Maps, Yandex.Music, Yandex.Image and Yandex.Video, among others. Yandex also offers location-based services like real-time road traffic congestion monitoring in the main cities through Yandex.Traffic, the most popular service of its kind in Russia. Yandex.Traffic was the first service in Russia, and one of the first in the world, to use GPS data from users' mobile devices (in anonymous form and with user consent) in assembling company's real-time congestion information. To extend its reach even further, Yandex designed a third service that enables its ads to be displayed on other, non-Yandex Web sites. This service, called Yandex ad network, allows third parties to earn a portion of Yandex's advertising revenue by showing Yandex ads on their own sites. Yandex ad network automatically analyzes a participating website's content and displays relevant text and image ads to visitors. Increased advertising earnings, in turn, motivate even more content owners to become Yandex ad network partners. Advertisers don't directly buy advertising space from Yandex. They bid on ad-related keywords associated with either search terms or content on third party websites. The bidding occurs through Yandex ad network auction service: the more popular a keyword, the more the advertiser has to pay for it.

The value proposition to the third party website owners, Yandex's third Customer Segment, is to enable them to earn money from their content. Adding new channels, Yandex offers downloadable applications for the most popular mobile and digital platforms and devices, and has introduced mobile versions of many of company's services. Yandex supports mobile phones, tablet devices, internet-connected TVs and navigation devices. Yandex's mobile apps include Yandex.Maps, Yandex.Search for Mobile, Yandex.Mail, Yandex.Money, Yandex.Market, Yandex.Taxi, Yandex.Fotki, Yandex.Metro and Yandex.Timetables.

Yandex SMSP's unique advantage is that it was created specifically for searching in Russian, considering Russian morphology and grammar. But let's look at what Yandex has to offer from the user's point of view.

At first glance, Yandex with its crowded homepage looks more like Yahoo rather than Google, with email, news, weather, picture of the day and all kinds of other features (http://www.yandex.ru). By the way, the dot-com version is much simpler (http://www.yandex.com/).

Among other unique features of Yandex are the following:

- **An Image Search Similar to That of Google: Yandex** offers a slideshow option, while Google allows you to start the search by dragging the image directly into the browser window;
- **Yandex Music:** A free online legal music player for users from Russia (feature of questionable utility for a country where illegal free download is the most popular way of obtaining music products, but anyway, Google doesn't have any similar service);

- **Yandex Money:** An e-commerce payment system convenient for online purchases, especially keeping in mind that Yandex operates in the part of the world where 'money' still means paper bills, not bank accounts;
- **Yandex Fotki:** A free photo-sharing service with unlimited storage space, similar to Yahoo's Flickr (there is no exact alternative on Google, but it can be compared to Google+);
- **Yandex.taxi, Yandex.Metro, Yandex.Local_Trains:** For Russia;
- **Traffic.ru:** A system for controlling online traffic;
- **Yandex Market and Yandex.Market App for Smartphones:** Personal shopping assistant that helps to find out comprehensive information on products specifications and prices in Russia;
- **Yandex News:** News, mostly from Russia;
- Advertising on Yandex is similar to advertising on Google;
- **Yandex Maps:** The undeniable advantage of Google Maps is global coverage. But consider the following: how often do you need to look at a map of a place other than the one you live or work in? The answer for most people would be 'not so often'. Yandex provides detailed maps of only 210 Russian and 57 Ukrainian cities (plus two cities in Kazakhstan and one in Belarus) – but those maps are equipped with some handy options that Google doesn't have. First of all, a great convenience for the Russian speakers is that all the writing on Yandex maps is in Cyrillic. Unlike Google, for big cities Yandex provides accurate directions for public transportation and allows its users to keep a track of the road traffic.

Yandex's key activities can be defined as follows:

- Building and maintaining the search infrastructure,
- Managing the three main services,
- Designing new services and products, and
- Promoting the platform to new users, content owners and advertisers.

Google SMSP has achieved the leadership in most countries virtually for free – simply because there were no national SMSPs there to start with. Wherever Google encountered some sort of resistance (such as China, the

Czech Republic, South Korea or Russia), it had to take a back seat. Analysis of Yandex SMSP's success on the Russian search market demonstrates that the company is reaping a strong focus on whatever is important for the local user, and that profound localization has been one of its key strengths, including local language support and a rich local content (such as maps, schedules, traffic, movies and news). It may, therefore, be concluded that this model might work on other local markets where Google has been formally localized at the source code page level, but continues to offer what is essentially no more than a simple query box.

CONCLUSION

The research framework for comparison and integration of different approaches for understanding search MSP strategic leadership and business models was developed in this work.

First we discussed the role of the leadership on a multi-sided market of search advertising, assuming the quantity competition and different entry conditions. Such approach can be developed taking into account network effects, multi-homing on the advertising side and the scale in search. If there are barriers to entry and the network effects are strong, there is an incentive for the leader to exploit them and attract more consumers to monopolize advertising. Under barriers to entry, the leading platform has also a strategic incentive to exploit the scale in search, to manipulate search results in order to divert search traffic from other platforms, and to introduce limits to multi-homing, with the aim of expanding its market share and deny scale to competitors.

Second, we proposed a definition of the search MSP and categorize the search MSP ecosystem taxonomy. The methodology, classification and organization of domain concepts, validation issues, and the first pilot version of search platform ontology/taxonomy and search platform business models were discussed.

Then a detailed search MSP's business model comparative analysis was developed for Google and Yandex SMSPs for Russian emerging search market. This research highlights the relevance of considering interdependences among the roles from emerging ecosystems (e.g., complementary products and services) when analyzing platform-based search markets. The case results show that regional search platform adoption in some countries is driven by the availability and use of regional complementary products/services. These findings reveal that decisions on complementary products/services in some cases act as strategic drivers for adoption in platform-based search markets.

REFERENCES

Etro, F. (2013). Advertising and search engines. A model of leadership in search advertising Original. *Research in Economics*, *67*(1), 25–38. doi:10.1016/j.rie.2012.10.001

Yablonsky, S. (2012). Yandex: The Russian Google (Case). The European Case Clearing House (ECCH), 2012, 912-041-1.

Yablonsky, S. (2016). Multi-sided search platforms: Global and local. *The International Journal of Technology Marketing*, *11*(3), 319–340. doi:10.1504/IJTMKT.2016.077394

Chapter 7
Integrated and Transactional Platforms:
Crowdsourcing and Crowdfunding Platforms

ABSTRACT

In recent years, crowdsourcing has increased in popularity as a method for gathering ideas for new innovations and providing solutions to existing problems. In this chapter, crowdsourcing intermediaries and business models are analyzed. How intermediaries' providers are managing interactions between different groups of platform actors (contributors) in order to receive new ideas, feedback, and solutions for improving consumers products and services is studied. Crowdfunding platforms are discussed based on crowdsourcing open innovation vision. Hence, research focuses on crowdfunding innovations for alternative financial services. The chapter aims to collect and analyze quality data regarding the current status and prospective evolution of crowdfunding. The study offers classification and examination of the current status of crowdfunding and proposes a definition of the crowdfunding multi-sided platform, develops research framework for crowdfunding platform comparison and business model analysis. The results of this chapter reveal benefits but also practical challenges to overcome before innovation crowdsourcing and crowdfunding intermediaries can be effectively utilized.

DOI: 10.4018/978-1-5225-5457-8.ch007

INTRODUCTION

The term "crowdsourcing" was introduced by Howe (2006) and defined as "the act of a company or institution taking a function once performed by employees and outsourcing it to an undefined (and generally large) network of people in the form of an open call. This can take the form of peer-production (when the job is performed collaboratively), but is also often undertaken by sole individuals. The crucial prerequisite is the use of the open call format and the large network of potential laborers."

Estellés-Arolas, E. and González-Ladrón-de-Guevara (2012) found more than 40 definitions of crowdsourcing in the extant literature. Based on those findings, they have created a comprehensive definition for crowdsourcing: "Crowdsourcing is a type of participative online activity in which an individual, an institution, a non-profit organization or company proposes to a group of individuals of varying knowledge, heterogeneity and number, via a flexible open call, the voluntary undertaking of a task. The undertaking of the task, of variable complexity and modularity, and in which the crowd should participate bringing their work, money, knowledge and/or experience, always entails mutual benefit. The user will receive the satisfaction of a given type of need, be it economic, social recognition, self- esteem or the development of individual skills, while the crowdsourcer will obtain and utilize to their advantage what the user has brought to the venture, whose form will depend on the type of activity undertaken."

Brabham (2008, 2013) defined "crowdsourcing" as an "online, distributed problem-solving and production model." Brabham defined the following 4 key ingredients of crowdsourcing:

1. An organization that has a task it needs to be performed,
2. A community (crowd) that is willing to perform the task voluntarily,
3. An online environment that allows the work to take place and the community to interact with the organization, and
4. Mutual benefit for the organization and the community.

What is more, crowdsourced labor is truly motivated to make a contribution only if they see their efforts make a real difference and the game is worth playing. That is why crowdsourcing is about networking and tapping the right crowd (Belleflamme et al., 2014).

Nakatsu, Grossman and Iacovou (2014) define crowdsourcing as a four-step process in which:

1. A requestor (either an individual or organization) identifies a specific task to be performed or problem to be solved;
2. The requestor broadcasts the task or problem online;
3. The crowd performs the task or solves the problem;
4. Depending on the nature of the task, the requestor either
    a. Sifts through the solutions and selects the best solution (selective crowdsourcing), or
    b. Aggregates/synthesizes the crowd's submissions in a meaningful way (integrative crowdsourcing)."

It is helpful to look at the levels of crowdsourcing classification (see Figure 1) to determine what is included in crowdsourcing approaches.

Crowdsourcing and Open Innovation

The open innovation concept was introduced by Chesbrough and Rosenbloom (2002), Chesbrough (2011). Marjanovic et al. (2012) and Wikhamn (2013) hold that crowdsourcing is an important type of open innovation.

While open innovation is a central idea in crowdsourcing, the two terms are not synonymous and should not be used interchangeably (Marjanovic et al., 2012). Open innovation focuses primarily on innovation processes, with many of these processes describing knowledge flows between firms (B2B transactions) rather than between firms and large crowds of anonymous individuals. As we have shown above, there are many other reasons why firms and individuals use crowdsourcing. So not all forms of open innovation are used for crowdsourcing (see Figure 2).

Crowdsourcing innovation is a multi-dimensional phenomenon that should be described from different perspectives – cultural, organizational and technological spheres – that have transformed significantly under the development of web technologies, web 2.0 in particular. A crowdsourcing process comprises a number of activities that involve resources within and beyond organizational boundaries, mostly human participants.

The characteristics of the activities and resources within this process depend on the specific purpose of a crowdsourcing effort. Research on crowdsourcing innovations comes from a variety of fields such as IT innovations, platform innovations, open innovations, social innovations, management, psychology and many other areas that have discovered crowdsourcing as a useful approach.

Crowdsourcing network users (seekers and solvers) access a common platform that facilitates their interactions. Platforms are usually a subset of

Figure 1. Classification of crowdsourcing approaches (2 levels)

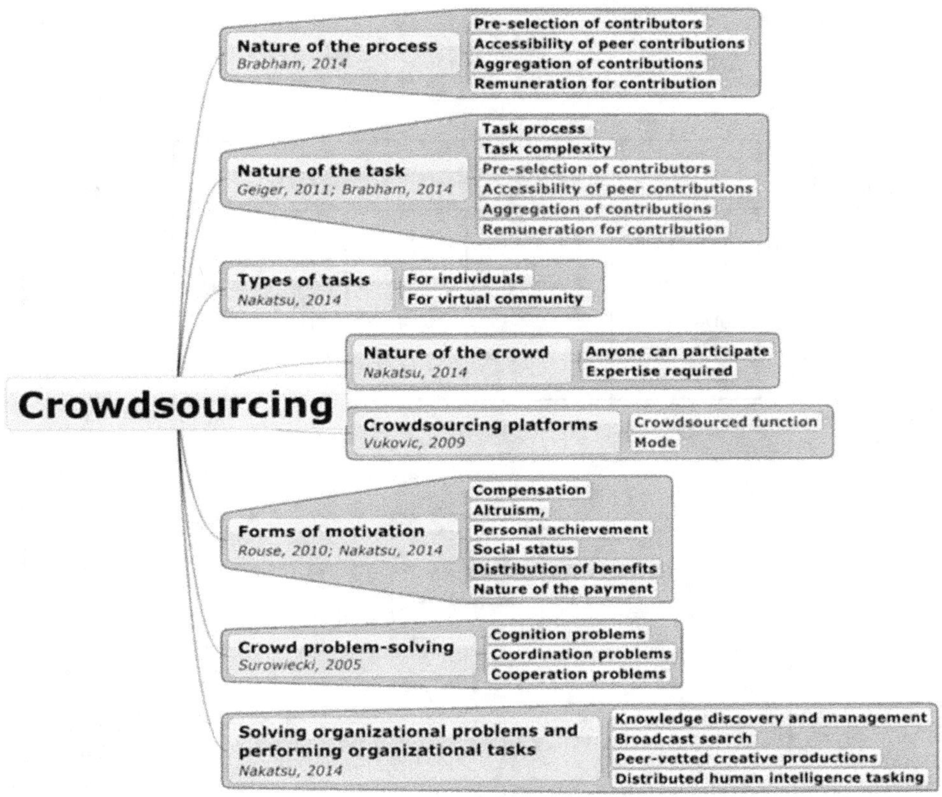

Figure 2. Classification of open innovation approaches (2 levels)

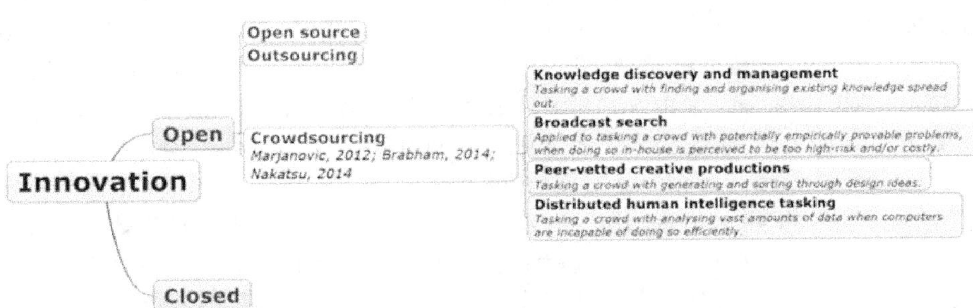

Figure 3. Three-sided crowdsourcing platform main building blocks

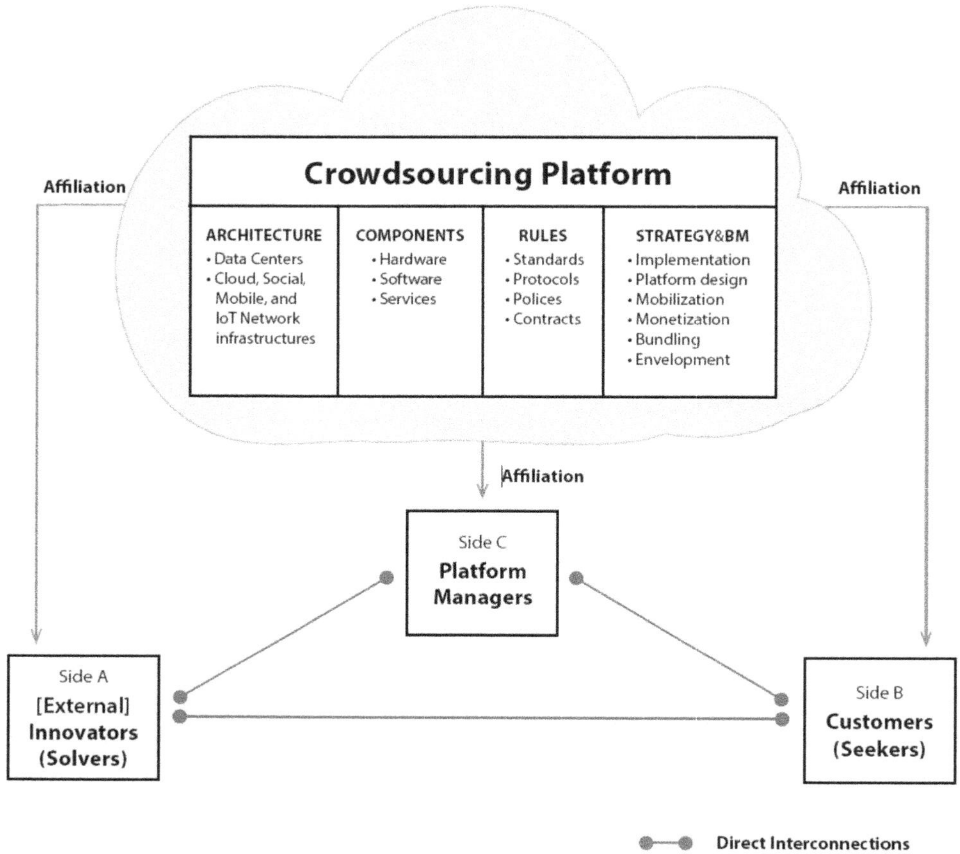

components (hardware, software, services) and rules (technical standards, protocols for information exchange, policies and contracts that govern transactions) employed by users in most of their transactions. Crowdsourcing MSP business model always should be related with Crowdsourcing MSP strategy. All this forms main platform building blocks (Figure 3). Minimal number of crowdsourcing MSP sides is usually 3.

Therefore, crowdsourcing MSP BMP framework has a distinct structure of taxonomy and is shown on the Figure 4. Crowdsourcing MSP BMP creates Value by facilitating interactions between different groups/sides of the platform. The crowdsourcing MSP BMP Value Proposition usually creates value in such main areas (Hagiu, 2006; Osterwalder & Pigneur, 2010; Evans, 2011):

Figure 4. Crowdsourcing MSP business model pattern

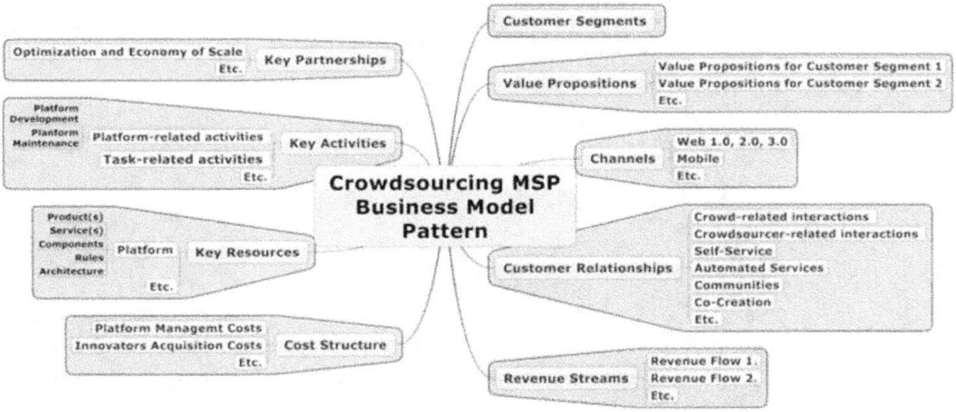

- Mobilizing user groups (Customer Segments) and balancing interests;
- Matchmaking between Customer Segments;
- Reducing costs by channeling transaction through the platform;
- Strategic decisions in platform evolution, multi-homing costs, scaling and liquidity.

Crowdsourcing MSP BMP has several Platform Customer Segments, each of which has its own Value Proposition and associated Revenue Stream. Furthermore in some business models one Customer Segment cannot exist without another. Sometimes one or more Customer Segments may enjoy free offers or reduced prices subsidized by revenues from other Customer Segments. Choosing which segment to subsidize can be a crucial strategic pricing decision that determines the success of MSP BM. MSPs are of value to one group of customers only if the other groups of customers are also present.

Crowdfunding Platforms

Crowdfunding is considered as an option or a part of crowdsourcing. In the recent years new forms of alternative finance including crowdfunding are growing quickly. This growth is beginning to attract institutional investors and alternative finance is on the cusp of becoming mainstream (Wardrop et al., 2015). Alternative financial services (AFS) are financial services provided inside or outside traditional financial institutions (Nesta, 2014; Wardrop

Figure 5. Alternative Finance Services (based on Wardrop et al., 2015)

et al., 2015). As it is mentioned by Wardrop et al. (2015), this innovative, decentralized and potentially disruptive alternative finances are supplying credit to consumers, providing early stage investments to start-ups and growth capital to SMEs, stimulating regional economies and funding worthwhile causes (Figure 5). Crowdfunding is a latest AFS innovation that appears to have tapped a new funding channel for entrepreneurs and enterprises. In particular, crowdfunding is emerging as an alternative funding channel for any stage of enterprise growth and becomes a powerful tool to drive innovations and bring the social change.

Despite its increased importance, limited research has systematically examined why and how non-financial and financial institutions engage in crowdfunding and how crowdfunding impacts alternative financial innovations. This chapter aims to address this critical gap by focusing on the following questions:

- What are the elements, relationships, vocabulary, and semantics of crowdfunding innovations?
- How to develop the methodology, the classification and organization of domain concepts of the crowdfunding innovation ontology?
- What are the basic types and the main business models of crowdfunding intermediaries?

We propose the crowdfunding ontology/taxonomy, which is a conceptualization and formalization of the elements, relationships, vocabulary and semantics of crowdfunding innovations. Our research was motivated by the observation that various descriptions of crowdfunding innovation services implement the concept of a crowdfunding services on similar core concepts with distinct features. The scope is the Global economy and the emerging economy, namely Russia. The goal is to capture the common core of different approaches to facilitate research in crowdfunding. Thus we illustrate how the crowdfunding innovation ontology can be complemented by the real instances in the Global market and emerging market of Russia.

As a result, the analysis methodology incorporates information from different but interrelated sources that form a representative sample of the domain:

- Reported and observed crowdfunding trends and e-finance activity. This study incorporates data from 2005 up to 2015 from the Internet (blogs, web conferences, etc.).
- Financial and non-financial organizations, finance/bank/crowdfunding cloud service providers, consulting companies, financial organizations and universities briefings, press releases, market and company reports, as well as other publicly available information.
- Research and text books in the field of crowdfunding, published from 2005 up to 2015.

A crowdfunding process comprises a number of activities that involve resources within and beyond organizational boundaries, mostly human participants and information technology (IT). The characteristics of the activities and resources within this process depend on the specific purpose of a crowdfunding effort. Research on crowdfunding innovations comes from a variety of fields such as IT innovations, platform innovations, open innovations, social innovations, management, psychology, and many other areas that have discovered crowdfunding as a useful approach. Accordingly, research questions mostly center on specific use cases and individual aspects of crowdfunding. This fragmentation of crowdfunding research and a general lack of foundational theory, however, make it difficult to provide organizations and entrepreneurs with well-founded guidance for the design of their crowdfunding initiatives (Haas, Blohm and Leimeister, 2014).

Schwienbacher and Larralde (2010) define crowdfunding as "an open call, essentially through the Internet, for the provision of financial resources either in form of donation or in exchange for some form of reward and/or voting rights in order to support initiatives for specific purposes." Mollick (2014) presents a definition in an entrepreneurial context: "Crowdfunding refers to the efforts by entrepreneurial individuals and groups – cultural, social and for profit – to fund their ventures by drawing on relatively small contributions from a relatively large number of individuals using the Internet, without standard financial intermediaries".

These definitions give the opportunity to associate to crowdfunding other financing modes like micro- finance and internet-based peer-to-peer lending and other cases.

Micro-finance concept means providing a limited financial support to low-income, unprivileged people that are non-secured from a usual banking point of view. The idea emerged recently and expanded rapidly around the world both online and offline. The best known platform for micro-financing abroad is Kiva, a non-profit organization that "provides safe, affordable access to capital to those in need and helps people create better lives for themselves and their families". Prosper is another well-known platform with similar purpose but whereas the former focuses the most on philanthropy the latter is more concentrated on transactional benefits and financial returns for both parties – "cutting out the middleman and connecting the people who need money with those who have it, borrowers get better rates and investors, better returns for their loans." While it is a wide-spread belief that micro-financing has a lot in common with crowdfunding, many authors refuse to accept this opinion – Caetano and Cardoso (2011) claims that "sites like Kiva and Prosper can be better placed within the finance sector than crowdfunding projects. That is because lenders charge interest rates."

Peer-to-peer lending relies on the same fundamentals as crowdfunding does but the applicability of these means of funds collection depends on the stage of business development. Peer-to-peer lending is a fast cash inflow to the business when no rewards need to be given away but the interest rate is paid on a constant basis just like it would happen in a bank. So it reminds mostly micro-financing process and has little to do with traditional crowdfunding.

Crowdfunding is considered as an option or a part of crowdsourcing. The term was introduced by Howe (2006) and defined as "the act of a company or institution taking a function once performed by employees and outsourcing it to an undefined (and generally large) network of people in the form of an open call. This can take the form of peer-production (when the job is performed collaboratively), but is also often undertaken by sole individuals. The crucial prerequisite is the use of the open call format and the large network of potential laborers." What is more, crowdsourced labor is truly motivated to make a contribution only if they see their efforts make a real difference and the game is worth playing. That is why crowdsourcing is about networking and tapping the right crowd, those being the points of parity with crowdfunding (Najah and Jarboui, 2015).

Crowdfunding Industry

Global Market

According to the statistical data for 2014, crowdfunding showed an excessive growth – crowdfunding platforms funded overall more than 2 million projects and collected more than $16,2 billion which makes a 162% of the figure for the same indicator in 2013 – $6,2 billion (Baeck, Collins and Zhang, 2014; Wardrop et al., 2015). Moreover, the crowdfunding industry is expected to double the results and reach $34,4 billion by the end of 2015. It may come as a surprise but the majority of funds – more than 53% – were collected on reward-free platforms that stand for only 30% of all crowdfunding platforms, while the most popular type – reward-based platforms – were less successful in raising funds despite of the fact they stand for 70% of all crowdfunding platforms. The total cumulative sum of investments in a reward-based crowdfunding rose by 549% annually, from $1,6 million in 2009. And it came as no surprise that the major part of investments is collected by Northern America ($837 million), which stands for more than a half of sum collected worldwide. North America has been the most rapidly growing region as well. By April 2014, there were 452 crowdfunding platforms worldwide, and the number was expected to grow by more than 600 by the end of 2015. Most of the platforms – 208 – are based in Northern America, 191 of which in USA. European region stands for 139 platforms, 44 of which are in Great Britain. Herewith, 95% of all funds attracted in Europe and 73% of all funds attracted in North America were assembled at the top five crowdfunding platforms. That happened mostly because of the size of the platforms, their reputation and expertise in audience knowledge, promotional activities and therefore – funds attraction.

The most well-known crowdfunding platforms are Kickstarter and its closest competitor Indiegogo. Other platforms have not reach yet such level of recognition and trust as those two.

Emerging Markets: Russia

Crowdfunding came to Russia in 2008 with a platform called Kroogi – it was launched by Miroslav Sabaev, who worked for a long time in American IT-companies and had an acquaintance with famous Russian musician Boris Grebenshikov. The idea of the project was the following: the musicians uploaded their musical works on the website and the fans were free to donate as much money as they wanted. Boris Grebenshikov was among the initiators of Kroogi campaign in Russia and uploaded a unique record from his concert which was not available anywhere else. Ultimately, a lot of musicians took part in the campaign and managed to earn as much as they would have probably earned working with record companies.

Another crowdfunding campaign that went aloud was RosPil activists oriented service organized by Alexey Navalny. Started in the middle of 2011, it gained huge popularity and quickly reached its target figures – 5 million rubles were collected within a month. That was some sort of signal for other political organizations – and if earlier activists were forced to fund their publishing and meetings by themselves, now this financial part could be shared with other sympathizers (Sokolov, 2015).

Yet politician projects are not able to attract as many investors from the crowd as projects devoted to charity, and the situation is quite clear – charity based on collective support existed long before the emergence of crowdfunding but the Internet just gave it a new wave, offering all the benefits of unlimited and boundless communication between parties of a campaign. In this connection, not only individuals and small non-profit organizations tend to apply to crowdfunding but also large holdings. Thus, in 2013 a federal TV Channel "One" launched a charity campaign for those people who were affected by the flood in the Far East of Russia and collected a significant amount of more than 700 million rubles.

Crowdfunding in Russia despite all those successful projects is still at the very beginning of its development. Experts estimate a total amount of

Russian crowdfunding market to be worth $2 million, which is incomparably smaller than European and American ones, but predict a 10 times grow in the future, though. Currently, there are already 9 working crowdfunding platforms. Only two of the existing Russian platforms have succeded so far: the biggest one is Planeta.ru with more than 50 million rubles collected so far, while the smaller one is Boomstarter with 10 million rubles raised. Some of other existing platforms are S miru po nitke, Tugeza and Kroogi.

Boomstarter is a Russian analog of the extremely popular American crowdfunding giant Kickstarter. In fact, it is the closest copy of the latter in everything – name, interface and functional opportunities. It was founded in 2012 by Ruslan Tugushev and Evgeny Gavrilin. Boomstarter helps authors of creative, scientific and other projects to attract funds by asking investors from the crowd. As a dividend, Boomstarter takes a 5% commission of all money collected, while another 5% goes to electronic paying systems. When placing a project to the platform, the author maintains all the rights concerned with property law but at the same time all project data goes public and cannot be deleted or somehow changed after the project's fulfillment. Projects that can be funded using the platform are classified by 12 categories, such as: design, food, publishing, fashion, technology, choreography, photography, theater, music, video and games, the last 3 ones being the most popular. Projects with a strictly determined aim like music album recording, book publishing or making a masterpiece, bounded with time frames and prototypes are welcomed as well, while charity, grant funding, medical and agitation campaigns are not allowed. Start-ups do not belong to the project classification so they are forbidden as well. Boomstarter applies the following scheme of collecting money: there is a strict rule that in order to transfer all money collected to the crowdfunder, the project needs to attract the whole sum that was initially stated during a certain period of time that is also determined beforehand. If the project does not manage to attract that sum in time, the money goes back to the investors.

Planeta.ru is another popular crowdfunding platform launched in 2012. Its system of money collection differs considerably from that of Boomstarter – no time boundaries and funding limits. If a crowdfunder manages to attract at least half of the sum initially stated, he can then take the money and start the project realization. The time span of the funding process is flexible as well – it is enough to inform the investors about a possible time period and that is it. Max Lakmus, the founder of Planeta.ru, points out that the restrictions on amounts and time applied by Boomstarter, prevent crowdfunding campaigns from successful fulfillment since the market for Russia is small and unstable so far. Besides, they do not reject charity projects and provide them with functional service without charging any commission. That advantageously distinguishes Planeta.ru from others and broadens its perspectives. By now Planeta.ru has 34 successfully completed and more than 45 on-going projects, it has helped to collect 9,5 million rubles and the number of registered users is dynamically approaching a 100 000 people border, more than 8 500 of whom have already participated in funding process.

Crowdfunding Regulations

Crowdfunding is developing rapidly around the world acting as an innovative mean of fundraising, but to facilitate effective transactions and ensure further growth it is important to establish a legislative base that would coordinate actions of all the parties of the process.

The history of crowdfunding legislative regulation in the US started on April 5, 2012, when the JOB Act was signed by the president Barak Obama. The Act implies a simplified system of investment collection for small-sized companies, so called start-ups without moving to IPO. The Act allows collecting up to $1 million with no more than 2 thousands investors as a maximum amount. In this case any restrictions concerned investors accreditations are removed. Each investor may participate in a crowdfunding campaign with a maximum amount of $10,000. Currently all crowdfunding platforms are available for accredited investors only.

It was stated, that during the thorough analysis of crowdfunding campaigns in the US, which reached a total of more than $250 million, no case of fraud was detected. Experts from Forbes believe that investments via crowdfunding platforms are more transparent and traceable in terms of accountability (Barnett, 2013).

The history of crowdfunding in Russia started in the second half of 2012 with the advent of such crowdfunding platforms as Boomstarter and Planeta.ru, which are currently developing very dynamically. And in spite of the fact that these projects were originally just copied from the American model, they have got their own characteristics, adapted to Russian realities. As a result, the explosive growth of interest to crowdfunding from domestic Internet users happened. Most Russian crowdfunding platform implement free-of-charge or conditionally repayable and conditionally return projects. In other words, there may be simple donations for a certain idea, or it may be a reward-based crowdfunding when investor gets either material or emotional value in return. From the point of view of Russian legislation such projects as crowdfunding fall under the regulation of art. 572 of the Civil Code and are seen as a donation contract. In this scheme, the investor, or more appropriately in this case – donor or giver, donates a certain amount of money to the recipient's property. The disadvantage of this structure is that it is actually excluded from legal entities by virtue of an outright ban on such transactions as well as the presence of additional constraints associated with the legislation to fight money laundering and maintain tax burden. One of the possible legal schemes that can be used to simplify the access of investors to start-ups, is available from July 1, 2014, after the entry into force of amendments to the Federal Law of 22.04.1996 N 39-FZ on the securities market concerning the legal status of the companies specialized on project financing. However, this form is only accessible for long-term (not less than 3 years) investment projects alongside with an issuance of bonds, which individuals can subsequently acquire.

So to sum up, there are no strict norms prohibiting from raising funds via crowdfunding but at the same time there are no norm supporting a simplified schemes of investment via the Internet either. And due to the fact that the essence of crowdfunding platforms lies in the possibility of becoming an investor remotely, without leaving home, it is advisable to consider some ways of simplifying the existing rules on the limited liability company (LLC) establishment. To implement crowdfunding projects within the LLC scheme it is necessary to remove the existing restriction on the number of participants and to simplify the procedure of re-registration of new members as well as the procedure of making important decisions. Taking into account all mentioned obstacles the development of a full-fledged investment in terms of crowdfunding in Russia would probably require the preparation of a special law, similar to the one signed in 2012 in the United States.

We propose the following definition of the crowdfunding multi-sided platform (Yablonsky, 2016b):

Crowdfunding multi-sided platform (henceforth, CMSP) is an organization that creates value primarily by enabling direct interactions between two main (or more) distinct types of affiliated customers: Capital- seeking Agents and Capital-giving Agents.

All this forms main crowdfunding multisided platform building blocks (Figure 6).

The study of CMSPs is important because it is a large and constantly growing share of global economy, and CMSPs' firms face distinctive management challenges: familiar rules, such as value-based pricing, may fail and traditional

Figure 6. Crowdfunding multisided platform building blocks

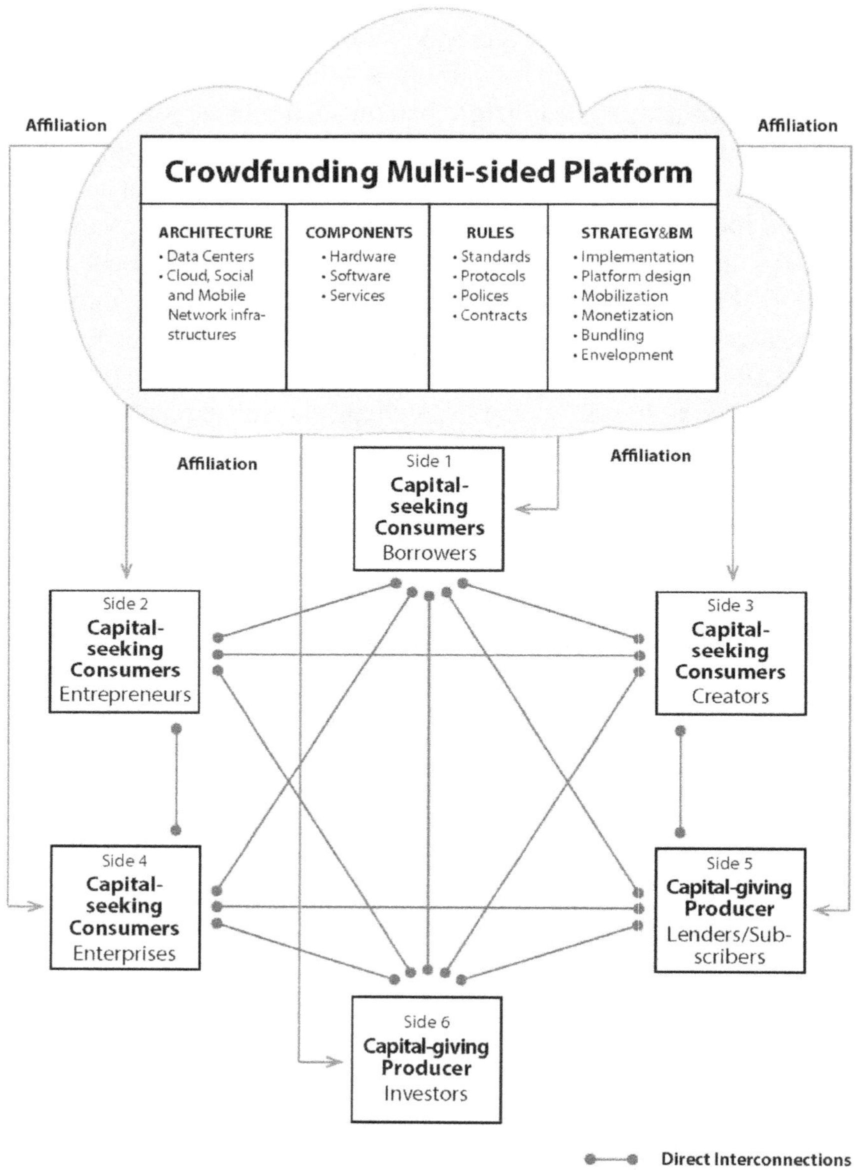

barriers to entry may no longer hold. Due to the network effects, platform intermediaries often enjoy increasing returns to scale and their industries have room for only a few players (Eisenmann, Parker, Van Alstyne, 2006).

Figure 7. Crowdfunding taxonomy

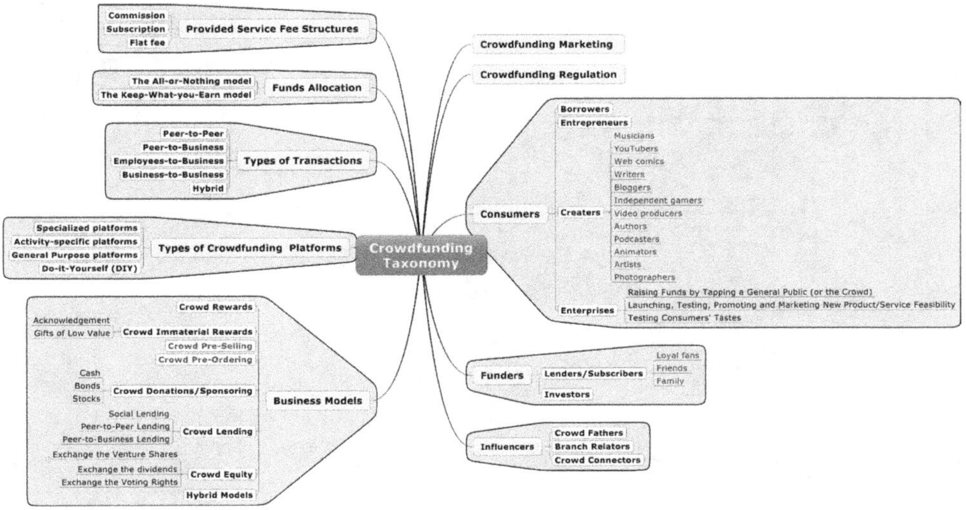

We propose crowdfunding ontology/taxonomy as a basic conceptual framework and knowledge management tool for analyzing and interpreting non-price instruments used by crowdfunding (Figure 7). General crowdfunding taxonomy suggests a framework for SMSP description and comparison.

We propose crowdfunding BMP (Figure 8). We set 7 main types of crowdfunding value propositions. The general crowdfunding business model sets the framework for thriving business model innovations that have been created and take into account the potential disruptive nature of crowdfunding. Going forward, more hybrid models will emerge. Some CMSP experiment with a combined model of loans and pre-sales. A percentage of the funding can be put into a loan (and will be repaid with interest) and the other part of the funding can be used to pre-finance the production of the product or service. Also there are benefits for the entrepreneur in hybrid models or in approaches to mix crowd-funding with other investment forms, for example, when crowdfunding is used to pre-sell a product, through which market validation and segmentation can be done to generate revenue and position the project for a follow-up or parallel investment from business angels. Crowdfunding is transformative in its total of business models, with the biggest potential lying in the combination of different approaches that will allow funding the whole life- cycle of a project, product, services or other business innovation.

In reality crowdfunding MSP use nuanced combinations of legal, technological, informational and other instruments (including price-setting) to implement desired outcomes (Boudreau and Hagiu, 2008).

Figure 8. Crowdfunding BMP

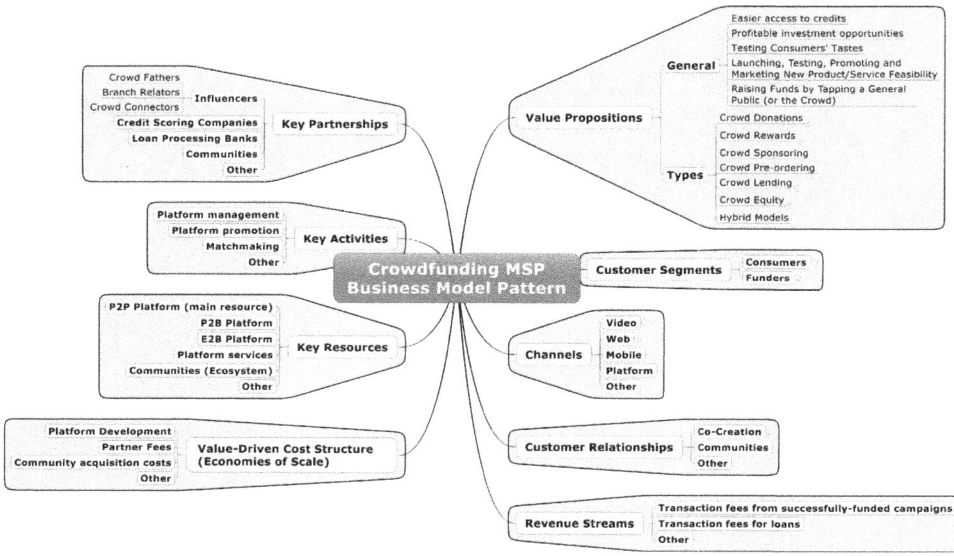

CONCLUSION

Summarizing we can conclude the following:

- Having analyzed innovation crowdsourcing e-intermediaries we propose a layered taxonomy framework for approaching the crowdsourcing in a platform business models context. The framework is illustrated by the example of crowdfunding and consulting platforms.
- A clear and precise description and structuring of the information on innovation crowdsourcing platform business models are prerequisites for a common understanding of the innovation crowdsourcing e-intermediaries. Following Osterwalder (2004) we provide an innovation crowdsourcing e-intermediaries business model pattern ontology, thT is a conceptualization and formalization of the elements, relationships, vocabulary and semantics of an e-intermediaries MSP (Yablonsky, 2016a). It is structured into several levels of decomposition with increasing depth and complexity. Ontologies, taxonomies and other types of controlled vocabularies are the preferred means to achieve such a common understanding by specifying the terms of the e-intermediary's domain, disambiguating them from each other, controlling synonyms and structuring the domain via term relationships.

- We propose a new definition of the innovation crowdsourcing multi-sided platform and the crowdfunding MSP.
- The research observes crowdfunding industry, including the global and the emerging markets (Russia), and proposes a definition of the crowdfunding multi-sided platform.
- The study offers a classification and examination of the current status of crowdfunding globally and in Russia.
- Crowdfunding is a term used to collectively describe many different delivery channels available to platform affiliated capital-seeking agents (investee) and capital-giving agents (investors) plus other members of crowdfunding ecosystem.
- The crowdfunding innovation taxonomy for global and Russian markets is deduced from the qualitative empirical research and offers a means to structure the understanding of the crowdfunding innovation services and main ecosystem actors, such as stakeholders, payment instruments, types, channels provided service fee structures, types of platforms, business models, agents, crowdfunding platforms, crowdfunding regulation, etc.
- Platform crowdfunding innovations prevail. Our research proposes a new definition of the crowdfunding multi-sided platform and develops a conceptual research framework for crowdfunding platform business models analysis, comparison and implementation.
- The general crowdfunding platform business model pattern sets the framework for thriving business model innovations that have been created and take into account the potential disruptive nature of crowdfunding. BMP contains 7 main types of crowdfunding value propositions.
- In the future more hybrid models will emerge. Some CMSP experiment with a combined model of loans and pre-sales. There are benefits for the entrepreneur in hybrid models or in approaches to mix crowd funding with other investment forms, for example where crowdfunding is used to pre-sell a product, through which market validation and segmentation can be done, to generate revenue and position the project for a follow-up or parallel investment from business angels.
- Platform crowdfunding is transformative in the total of its business models, with the biggest potential lying in the combination of different approaches that will allow funding the whole life cycle of a project, product, service or other business innovation.

- Crowdfunding platform is opposed by the prevalent information asymmetry between both parties. The investee holds all the information while the investor has only limited chances to verify the truth and relevance of the information given by the platform.
- The main limitations of crowdfunding innovations are related with the lack of a unified regulation between the US, Europe and the Emerging countries.
- *Practical implications.* The evaluation logic presents a tool that managers, company owners and investors can use to identify innovation crowdsourcing business models to interact with different groups of contributors in order to receive new ideas, feedback and solutions, so that it can be later used to evaluate the effectiveness of innovation crowdsourcing platforms and in decision-making regarding the design of the value offering of platform's products and services. Since the framework has been specifically designed for evaluating the innovation crowdsourcing potential of platforms, it proves to be particularly convenient for analyzing different types of platforms. It can be used not only to evaluate the effectiveness of innovation crowdsourcing platforms, but also in decision-making concerning the design of a different value offering of innovation crowdsourcing multi-sided platform business models and products or services.

REFERENCES

Baeck, P., Collins, L., & Zhang, B. (2014). *Undertaking Alternative Finance.* The UK Alternative Finance Industry Report. Retrieved December 18, 2017, from: https://www.nesta.org.uk/sites/default/files/understanding-alternative-finance-2014.pdf

Barnett, C. (2013). *Donation-based crowdfunding sites: Kickstarter vs. Indiegogo.* Retrieved December 18, 2017, from: http://www.forbes.com/sites/chancebarnett/2013/09/09/donation-based- crowdfunding-sites-kickstarter-vs-indiegogo/

Belleflamme, P., Lambert, T., & Schwienbacher, A. (2014). Crowdfunding: Tapping the right crowd. *Journal of Business Venturing*, *29*(5), 585–609. doi:10.1016/j.jbusvent.2013.07.003

Boudreau, K. J., & Hagiu, A. (2008). *Platform rules: Multi-sided platforms as regulators*. Available at SSRN 1269966

Brabham, D. (2008). Crowdsourcing as a Model for Problem Solving: An Introduction and Cases. *Convergence (London), 14*(1), 75–90. doi:10.1177/1354856507084420

Brabham, D. (2013). *Crowdsourcing*. MIT Press.

Caetano, M. A., & Cardoso, G. (2011). The Democratization of film financing: A comparative analysis of crowdfunding sites and projects for independent movies. *Marketing Letters, 12*(3), 222–237.

Eisenmann, T., Parker, G., & Van Alstyne, M. (2006). Strategies for Two-Sided Markets. *Harvard Business Review*, 10.

Estellés-Arolas, E., & González-Ladrón-de-Guevara, F. (2012). Towards an integrated crowdsourcing definition. *Journal of Information Science*, *38*(2), 189–200. doi:10.1177/0165551512437638

Evans, D. S., Hagiu, A., & Schmalensee, R. (2006). *Invisible engines: how software platforms drive innovation and transform industries*. Cambridge, MA: MIT Press.

Evans, D. S., Schmalensee, R., Noel, M. D., Chang, H. H., & Garcia-Swartz, D. D. (2011). Platform economics: Essays on multi-sided businesses. Platform Economics: Essay on Multi-sided Businesses. *Competition Policy International*.

Haas, P., Blohm, I., & Leimeister, J. (2014). An Empirical Taxonomy of Crowdfunding Intermediaries. *Proceedings Thirty Fifth International Conference on Information Systems*.

Howe, J. (2006). The Rise of crowdsourcing. *Wired Magazine*, *14*(6), 34–36.

Marjanovic, S., Fry, C., & Chataway, J. (2012). Crowdsourcing based business models: In search of evidence for innovation 2.0. *Science & Public Policy*, *39*(3), 1–15. doi:10.1093cipolcs009

Mollick, E. (2014). The dynamics of crowdfunding: An exploratory study. *Journal of Business Venturing*, *29*(1), 1–16. doi:10.1016/j.jbusvent.2013.06.005

Najah, A., & Jarboui, A. (2015). Crowd funding for financing innovative and social entrepreneurship: Literature Review. *Journal of Business Management and Economics*, *3*(6).

Nakatsu, R., Grossman, E., & Iacovou, C. (2014). A taxonomy of crowdsourcing based on task complexity. *Journal of Information Science*, *40*(6), 823–834. doi:10.1177/0165551514550140

Nesta. (2014). *Understanding Alternative Finance.* The UK Alternative Finance industry Report 2014. Retrieved December 18, 2017, from: https://www.nesta.org.uk/sites/default/files/understanding-alternative- finance-2014.pdf

Osterwalder, A. (2004). *The Business Model Ontology - A proposition in a design science approach* (Ph.D. thesis). University of Lausanne, Ecole des Hautes Etudes Commerciales HEC.

Osterwalder, A., & Pigneur, I. (2010). *Business Model Generation: A Handbook for Visionaries, Game Changers, and Challengers.* John Wiley & Sons.

Schwienbacher, A., & Larralde, B. (2010). Crowdfunding of small entrepreneurial ventures. In *Handbook of Interpreneurial Finance*. Oxford University Press. doi:10.2139srn.1699183

Social Innovation Lab. (2013). *Social Business Model Canvas.* Retrieved July 20, 2015, from http://www.socialbusinessmodelcanvas.com

Sokolov, A. (2015). Russian Political Crowdfunding, Demokratizatsiya. *The Journal of Post-Soviet Democratization*, *23*(2), 117–149.

Wardrop, R., Zhang, B., Rau, R., & Gray, M. (2015). *Moving Mainstream. The European Alternative Finance Benchmarking Report*. Retrieved December 18, 2017, from: http://capacity4dev.ec.europa.eu/crowdfunding/document/european-alternative-finance-benchmarking-report

Wikhamn, B. (2013). Two Different Perspectives on Open Innovation. *Creativity and Innovation Management*, *22*(4), 375–389. doi:10.1111/caim.12035

Yablonsky, S. (2016a). Intermediaries in E-Commerce. In *Encyclopedia of E-Commerce Development, Implementation, and Management* (pp. 48–73). IGI Global. doi:10.4018/978-1-4666-9787-4.ch005

Yablonsky, S. (2016b). Crowdfunding Innovations. *International Journal of Services, Economics and Management*, *7*(2/3/4). doi:https://doi.org/10.1504/IJSEM.2016.081870 doi:10.1504/IJBEX.2017.082587

Conclusion

In reply to the challenges of the digital economy we need more than just tools for the development of a digital technology.

This book has reviewed the following key aspects of platform research: *platform strategy, dynamic capabilities and business models*. The main *platform typologies and basic definitions* were described first. This research has attempted to provide a literature review relating to arguing platform strategy, platform vue chain, platform life cycle, platform building blocks and business models.

It went on by suggesting a *platform strategy categorization taxonomy* and *platform practical strategies of preventing platform disintermediation*. Research on platform business models is on the rise, and this book has argued that platform business model analysis and design could be explained by categorization of the *platform business model building blocks taxonomy* that was supported by strategy, dynamic capabilities, *platform business models framework* and *business model innovation process framework*. The *main types of platform business models* were introduced. Selecting the platform business model and the platform strategy could reformulate more precisely what type of business the owner of the platform technology wants to become.

Business Model Patterns concept is another way of structuring digital value drivers (Gassmann et al. 2014a, 2014b, 2016). Following Osterwalder (2004) and Osterwalder et al. (2010) the *multi-sided platform business model pattern (MSP BMP) framework* was introduced and discussed. MSP BMP is provided as a basic conceptual framework and knowledge management tool for describing, analyzing and interpreting non-price instruments used by digital platforms, especially platform intermediaries. Finally, platform business model design methodology was introduced. The book moves on to consider the 12 dimensions of the BMP taxonomy for defining MSP BMP dimensions that could be used as a platform assessment tool. Also the proposed methodology could be used for systematic Digital Business Modelling based on a language both business and technology experts equally understand. MSP BMP framework with meta-components could be used in research of platform identification/creation/assessment/disclosure.

In the context of digital business, a platform usually supports and facilitates sharing of key platform resources/assets with ecosystems of people, businesses and things. A platform typically includes tools, functions and support for multiple customer segments (e.g., buyers and sellers; students, faculty and administrators).

This research focused on the *Digital platform stack*: necessary platform components like technology, leadership, talent and skills, delivery, trust, marketing, ecosystem and business models, required to support the capabilities of digital business innovations. The multidimensional platform organization and management was discussed.

Using the MSP BMP as the basis of the digital intermediaries' analysis and decision making, this pattern was updated for different industries and technological innovations in chapters of the book. The *platform business model patterns* highlight a number of specific design ideas, and they detail them using companies having successfully implemented them as examples. Platform Business Model Pattern dimensions framework with meta-components could be used in identification, creation, assessment and disclosure research of platform businesses.

As it was mentioned (Blaschke et al., 2017a, 2017b) what we need is "a methodology for systematic Digital Business Modelling based on a language both business and technology experts equally understand". The structured approach that was presented in this book is a step in this direction.

REFERENCES

Blaschke, M., Haki, M., Riss, U., & Aier, S. (2017a). Design Principles for Business-Model-based Management Methods—A Service-dominant Logic Perspective. *Proceedings International Conference on Design Science Research in Information Systems and Technology (DESRIST)*. doi:10.1007/978-3-319-59144-5_11

Blaschke, M., Cigaina, M., Riss, U., & Shoshan, I. (2017b). Designing Business Models for the Digital Economy. In G. Oswald & M. Kleinemeier (Eds.), *Shaping the Digital Enterprise. Trends and Use Cases in Digital Innovation and Transformation*. Springer International Publishing Switzerland. doi:10.1007/978-3-319-40967-2_6

Gassmann, O., Frankenberger, K., & Csik, M. (2014a). *The Business Model Navigator: 55 Models That Will Revolutionise Your Business*. Pearson Education Limited.

Gassmann, O., Frankenberger, K., & Csik, M. (2014b). Revolutionizing the business model. In O. Gassmann & F. Schweitzer (Eds.), *Management of the fuzzy front end of innovation*. Springer International Publishing. doi:10.1007/978-3-319-01056-4_7

Gassmann, O., Frankenberger, K., & Sauer, R. (2016). *Exploring the Field of Business Model Innovation. New Theoretical Perspectives*. Palgrave Macmillan. doi:10.1007/978-3-319-41144-6

Osterwalder, A. (2004). *The Business Model Ontology - A proposition in a design science approach* (Ph.D. thesis). University of Lausanne, Ecole des Hautes Etudes Commerciales HEC.

Osterwalder, A., & Pigneur, I. (2010). *Business Model Generation: A Handbook for Visionaries, Game Changers, and Challengers*. John Wiley & Sons.

Related Readings

To continue IGI Global's long-standing tradition of advancing innovation through emerging research, please find below a compiled list of recommended IGI Global book chapters and journal articles in the areas of economics, e-business, and economic policies. These related readings will provide additional information and guidance to further enrich your knowledge and assist you with your own research.

Abereijo, I. O. (2016). Ensuring Environmental Sustainability through Sustainable Entrepreneurship. In A. Goswami & A. Mishra (Eds.), *Economic Modeling, Analysis, and Policy for Sustainability* (pp. 234–249). Hershey, PA: IGI Global. doi:10.4018/978-1-5225-0094-0.ch012

Acosta, Y. A., Zapata, I. T., & Dornberger, U. (2014). Technology-Intensive Suppliers as a Key Element for Structural Change in Latin America. In B. Christiansen (Ed.), *Handbook of Research on Economic Growth and Technological Change in Latin America* (pp. 44–54). Hershey, PA: IGI Global. doi:10.4018/978-1-4666-6224-7.ch003

Adams, S., Klobodu, E. K., & Lamptey, R. O. (2017). Electric Power Transmission, Distribution Losses, and Economic Growth in Ghana. In R. Das (Ed.), *Social, Health, and Environmental Infrastructures for Economic Growth* (pp. 216–233). Hershey, PA: IGI Global. doi:10.4018/978-1-5225-2364-2.ch011

Adams, S., Klobodu, E. K., & Lamptey, R. O. (2017). Health Infrastructure and Economic Growth in Sub-Saharan Africa. In R. Das (Ed.), *Social, Health, and Environmental Infrastructures for Economic Growth* (pp. 82–98). Hershey, PA: IGI Global. doi:10.4018/978-1-5225-2364-2.ch005

Adetiloye, K. A., Eke, P. O., & Taiwo, J. N. (2017). Public-Sector Project Abandonment Decision: A Test of the Ricardian Equivalence Theory on the Failed Lagos Metroline in Nigeria. In R. Das (Ed.), *Social, Health, and Environmental Infrastructures for Economic Growth* (pp. 188–215). Hershey, PA: IGI Global. doi:10.4018/978-1-5225-2364-2.ch010

Al-Somali, S. A., Gholami, R., & Clegg, B. (2015). An Investigation into the Factors Affecting E-Commerce Adoption Decisions by SMEs: A Study in Saudi Arabia. In M. Khosrow-Pour (Ed.), *Strategic E-Commerce Systems and Tools for Competing in the Digital Marketplace* (pp. 206–243). Hershey, PA: IGI Global. doi:10.4018/978-1-4666-8133-0.ch011

Al-Subhi, A. A., & Alfares, H. K. (2016). Economic Load Dispatch Using Linear Programming: A Comparative Study. *International Journal of Applied Industrial Engineering, 3*(1), 16–36. doi:10.4018/IJAIE.2016010102

Albeck, W., & Huth, S. (2014). The Internationalization Process of Multilatinas from Chile. In B. Christiansen (Ed.), *Handbook of Research on Economic Growth and Technological Change in Latin America* (pp. 209–229). Hershey, PA: IGI Global. doi:10.4018/978-1-4666-6224-7.ch012

Alderete, M. V. (2014). Internet Incidence on SME's Sales: A Propensity Score Matching Analysis. In I. Lee (Ed.), *Trends in E-Business, E-Services, and E-Commerce: Impact of Technology on Goods, Services, and Business Transactions* (pp. 175–190). Hershey, PA: IGI Global. doi:10.4018/978-1-4666-4510-3.ch010

Alfahl, H., Sanzogni, L., Houghton, L., & Sandhu, K. (2014). Mobile Commerce Adoption in Organizations: A Literature Review and Preliminary Findings. In I. Lee (Ed.), *Trends in E-Business, E-Services, and E-Commerce: Impact of Technology on Goods, Services, and Business Transactions* (pp. 47–68). Hershey, PA: IGI Global. doi:10.4018/978-1-4666-4510-3.ch003

Allard, G. (2014). A Different Kind of Investor?: Chinese FDI in Latin America. In B. Christiansen (Ed.), *Handbook of Research on Economic Growth and Technological Change in Latin America* (pp. 417–433). Hershey, PA: IGI Global. doi:10.4018/978-1-4666-6224-7.ch024

Allevato, E., & Marques, J. (2014). Brazil: Economic Mirage or Jewel in the New Global Economy? In B. Christiansen (Ed.), *Handbook of Research on Economic Growth and Technological Change in Latin America* (pp. 55–72). Hershey, PA: IGI Global. doi:10.4018/978-1-4666-6224-7.ch004

Amone, W. (2016). Game Theory. In B. Christiansen & E. Lechman (Eds.), *Neuroeconomics and the Decision-Making Process* (pp. 262–286). Hershey, PA: IGI Global. doi:10.4018/978-1-4666-9989-2.ch014

Angelini, P. P. (2016). The Role of Proximity in Inter-Organizational Network Evolution: An Application of Actor-Oriented Modeling for Longitudinal Network Analysis. In L. Biggiero, P. Angelini, M. Basevi, N. Carbonara, A. Mastrogiorgio, E. Pessa, ... M. Valente (Eds.), *Relational Methodologies and Epistemology in Economics and Management Sciences* (pp. 232–256). Hershey, PA: IGI Global. doi:10.4018/978-1-4666-9770-6.ch008

Angelini, P. P., & Biggiero, L. (2016). A Comparison between International Trade and R&D Collaboration Networks in the European Aerospace Sector. In L. Biggiero, P. Angelini, M. Basevi, N. Carbonara, A. Mastrogiorgio, E. Pessa, ... M. Valente (Eds.), *Relational Methodologies and Epistemology in Economics and Management Sciences* (pp. 141–171). Hershey, PA: IGI Global. doi:10.4018/978-1-4666-9770-6.ch005

Artienwicz, N. (2016). Behavioral Stream in Polish Accounting: Its Relation to Behavioral Finance and the Perspectives for Neuroaccounting Development in Poland. In B. Christiansen & E. Lechman (Eds.), *Neuroeconomics and the Decision-Making Process* (pp. 246–261). Hershey, PA: IGI Global. doi:10.4018/978-1-4666-9989-2.ch013

Atli, D., & Yilmazata, M. (2016). Neuroeconomics and Media Economics. In B. Christiansen & E. Lechman (Eds.), *Neuroeconomics and the Decision-Making Process* (pp. 33–44). Hershey, PA: IGI Global. doi:10.4018/978-1-4666-9989-2.ch002

Ayuk, E. T., Fonta, W. M., & Kouame, E. B. (2014). Application of Quantitative Methods in Natural Resource Management in Africa: A Review. In P. Schaeffer & E. Kouassi (Eds.), *Econometric Methods for Analyzing Economic Development* (pp. 205–234). Hershey, PA: IGI Global. doi:10.4018/978-1-4666-4329-1.ch013

Bachrane, M., El Alami, J., & Hanoune, M. (2015). Economic Intelligence Information and Process in Shipping Companies in Morocco: The Case of TangerMed Area. *Journal of Information Technology Research*, *8*(2), 50–60. doi:10.4018/jitr.2015040104

Banerji, D., & Das, R. (2014). Understanding the Impact of Securities Markets Reform on the Economy of Brazil. In B. Christiansen (Ed.), *Handbook of Research on Economic Growth and Technological Change in Latin America* (pp. 339–349). Hershey, PA: IGI Global. doi:10.4018/978-1-4666-6224-7.ch019

Barat, S., & Spillan, J. E. (2014). An Exploratory Study of Customer Satisfaction in a Community Bank. In I. Lee (Ed.), *Trends in E-Business, E-Services, and E-Commerce: Impact of Technology on Goods, Services, and Business Transactions* (pp. 111–127). Hershey, PA: IGI Global. doi:10.4018/978-1-4666-4510-3.ch006

Basevi, M., & Biggiero, L. (2016). An International Trade Comparison of Two Supposedly Different Sectors: An Investigation on Inter-Sectoral Diversity and Sectoral Trade Stability. In L. Biggiero, P. Angelini, M. Basevi, N. Carbonara, A. Mastrogiorgio, E. Pessa, ... M. Valente (Eds.), *Relational Methodologies and Epistemology in Economics and Management Sciences* (pp. 172–205). Hershey, PA: IGI Global. doi:10.4018/978-1-4666-9770-6.ch006

Basılgan, M., & Christiansen, B. (2014). Taxpayers' Attitudes towards Tax Evasion in Latin American Countries. In B. Christiansen (Ed.), *Handbook of Research on Economic Growth and Technological Change in Latin America* (pp. 73–104). Hershey, PA: IGI Global. doi:10.4018/978-1-4666-6224-7.ch005

Bassi, A., & Gallagher, L. (2016). Integrated Economic and Spatial Planning for the Food-Energy-Water Nexus. In A. Goswami & A. Mishra (Eds.), *Economic Modeling, Analysis, and Policy for Sustainability* (pp. 54–73). Hershey, PA: IGI Global. doi:10.4018/978-1-5225-0094-0.ch005

Basu, I. (2017). Role of Infrastructure Development to Empower Women: An Over-Determined View. In R. Das (Ed.), *Social, Health, and Environmental Infrastructures for Economic Growth* (pp. 39–56). Hershey, PA: IGI Global. doi:10.4018/978-1-5225-2364-2.ch003

Bhattacharyya, S. C., & Palit, D. (2016). Enabling Policies for Advancing Sustainability of Electricity Access Programs. In A. Goswami & A. Mishra (Eds.), *Economic Modeling, Analysis, and Policy for Sustainability* (pp. 177–193). Hershey, PA: IGI Global. doi:10.4018/978-1-5225-0094-0.ch009

Bieri, D. S. (2014). Regional Structure and Economic Development: Growth Empirics for U.S. Metropolitan Areas. In P. Schaeffer & E. Kouassi (Eds.), *Econometric Methods for Analyzing Economic Development* (pp. 153–176). Hershey, PA: IGI Global. doi:10.4018/978-1-4666-4329-1.ch010

Biggiero, L. (2016). Network Analysis for Economics and Management Studies. In L. Biggiero, P. Angelini, M. Basevi, N. Carbonara, A. Mastrogiorgio, E. Pessa, ... M. Valente (Eds.), *Relational Methodologies and Epistemology in Economics and Management Sciences* (pp. 1–60). Hershey, PA: IGI Global. doi:10.4018/978-1-4666-9770-6.ch001

Biggiero, L. (2016). NK Simulation Modeling. In L. Biggiero, P. Angelini, M. Basevi, N. Carbonara, A. Mastrogiorgio, E. Pessa, ... M. Valente (Eds.), *Relational Methodologies and Epistemology in Economics and Management Sciences* (pp. 61–99). Hershey, PA: IGI Global. doi:10.4018/978-1-4666-9770-6.ch002

Biggiero, L., & Basevi, M. (2016). Knowledge Creation, Growth, and Transfer within Industrial Networks of Practices: The Role of Absorptive Capacity and Direct Centrality. In L. Biggiero, P. Angelini, M. Basevi, N. Carbonara, A. Mastrogiorgio, E. Pessa, ... M. Valente (Eds.), *Relational Methodologies and Epistemology in Economics and Management Sciences* (pp. 313–347). Hershey, PA: IGI Global. doi:10.4018/978-1-4666-9770-6.ch012

Biggiero, L., & Mastrogiorgio, A. (2016). A Methodology to Measure the Hierarchical Degree of Formal Organizations. In L. Biggiero, P. Angelini, M. Basevi, N. Carbonara, A. Mastrogiorgio, E. Pessa, ... M. Valente (Eds.), *Relational Methodologies and Epistemology in Economics and Management Sciences* (pp. 206–231). Hershey, PA: IGI Global. doi:10.4018/978-1-4666-9770-6.ch007

Biggiero, L., & Sevi, E. (2016). On the Relationships between Connection Modes and Workgroup Performance: The Moderating Role of Group Size and Task Complexity. In L. Biggiero, P. Angelini, M. Basevi, N. Carbonara, A. Mastrogiorgio, E. Pessa, ... M. Valente (Eds.), *Relational Methodologies and Epistemology in Economics and Management Sciences* (pp. 348–372). Hershey, PA: IGI Global. doi:10.4018/978-1-4666-9770-6.ch013

Biggiero, L., & Valente, M. (2016). The Role of Knowledge Heterogeneity in Industrial Clusters Knowledge Dynamics: An Application of the Boolean Network Simulation Modeling. In L. Biggiero, P. Angelini, M. Basevi, N. Carbonara, A. Mastrogiorgio, E. Pessa, ... M. Valente (Eds.), *Relational Methodologies and Epistemology in Economics and Management Sciences* (pp. 257–282). Hershey, PA: IGI Global. doi:10.4018/978-1-4666-9770-6.ch009

Bishop, J. (2014). Reducing Corruption and Protecting Privacy in Emerging Economies: The Potential of Neuroeconomic Gamification and Western Media Regulation in Trust Building and Economic Growth. In B. Christiansen & M. Basilgan (Eds.), *Economic Behavior, Game Theory, and Technology in Emerging Markets* (pp. 237–249). Hershey, PA: IGI Global. doi:10.4018/978-1-4666-4745-9.ch013

Bouras, C., Antonis, K., Diles, G., Kokkinos, V., & Loukopoulos, L. (2014). Economic Broadband Development through Infrastructure Sharing. *International Journal of Business Data Communications and Networking*, *10*(4), 21–42. doi:10.4018/IJBDCN.2014100102

Bravo, R., Hem, L. E., & Pina, J. M. (2014). Online Brand Expansion towards the Offline Setting: Which Way to Go? In I. Lee (Ed.), *Trends in E-Business, E-Services, and E-Commerce: Impact of Technology on Goods, Services, and Business Transactions* (pp. 69–89). Hershey, PA: IGI Global. doi:10.4018/978-1-4666-4510-3.ch004

Buechler, S., Hartshorne, R., & Ajjan, H. (2014). Examining the Impact of Web 2.0 Applications on Knowledge Management Performance. In I. Lee (Ed.), *Trends in E-Business, E-Services, and E-Commerce: Impact of Technology on Goods, Services, and Business Transactions* (pp. 90–110). Hershey, PA: IGI Global. doi:10.4018/978-1-4666-4510-3.ch005

Buğa, D. (2016). Central Language Hypothesis in the Decision-Making Process. In B. Christiansen & E. Lechman (Eds.), *Neuroeconomics and the Decision-Making Process* (pp. 66–83). Hershey, PA: IGI Global. doi:10.4018/978-1-4666-9989-2.ch004

Burger, R., & du Plessis, S. (2014). A New Keynesian Phillips Curve for South Africa. In P. Schaeffer & E. Kouassi (Eds.), *Econometric Methods for Analyzing Economic Development* (pp. 30–48). Hershey, PA: IGI Global. doi:10.4018/978-1-4666-4329-1.ch003

Byambasuren, T., Purevjav, A., & Erdenekhuyag, E. (2015). Economic Impacts of the Government Investment Policy: Dynamic CGE Model. *International Journal of System Dynamics Applications*, *4*(1), 96–118. doi:10.4018/IJSDA.2015010104

Carbonara, N. (2016). Knowledge-Driven Agglomeration Processes: An Agent-Based Simulation Study. In L. Biggiero, P. Angelini, M. Basevi, N. Carbonara, A. Mastrogiorgio, E. Pessa, ... M. Valente (Eds.), *Relational Methodologies and Epistemology in Economics and Management Sciences* (pp. 373–396). Hershey, PA: IGI Global. doi:10.4018/978-1-4666-9770-6.ch014

Cartwright, E., & Singh, T. (2014). Social Capital, the Culture of Trust, and Economic Development. In B. Christiansen & M. Basilgan (Eds.), *Economic Behavior, Game Theory, and Technology in Emerging Markets* (pp. 91–108). Hershey, PA: IGI Global. doi:10.4018/978-1-4666-4745-9.ch006

Chandan, H. C. (2014). Business Ethics in Latin America and Its Impact on Sustained Economic Growth. In B. Christiansen (Ed.), *Handbook of Research on Economic Growth and Technological Change in Latin America* (pp. 154–173). Hershey, PA: IGI Global. doi:10.4018/978-1-4666-6224-7.ch009

Chandan, H. C. (2014). The Role of Religion and National Culture in Economic Growth of Emerging Markets. In B. Christiansen & M. Basilgan (Eds.), *Economic Behavior, Game Theory, and Technology in Emerging Markets* (pp. 250–270). Hershey, PA: IGI Global. doi:10.4018/978-1-4666-4745-9.ch014

Chandan, H. C. (2016). Rational, Emotional, and Neural Foundations of Economic Preferences. In B. Christiansen & E. Lechman (Eds.), *Neuroeconomics and the Decision-Making Process* (pp. 45–65). Hershey, PA: IGI Global. doi:10.4018/978-1-4666-9989-2.ch003

Chang, V. (2014). A Case Study for Business Integration as a Service. In I. Lee (Ed.), *Trends in E-Business, E-Services, and E-Commerce: Impact of Technology on Goods, Services, and Business Transactions* (pp. 229–254). Hershey, PA: IGI Global. doi:10.4018/978-1-4666-4510-3.ch013

Chatterjee, N. (2017). An Assessment of Infrastructural Facilities in the Dryland Areas of West Bengal. In R. Das (Ed.), *Social, Health, and Environmental Infrastructures for Economic Growth* (pp. 234–268). Hershey, PA: IGI Global. doi:10.4018/978-1-5225-2364-2.ch012

Chen, Y., Watson, E., & Azevedo, R. F. (2014). E-Strategy and Soft Landings for Franchising in Emerging Markets. In I. Lee (Ed.), *Trends in E-Business, E-Services, and E-Commerce: Impact of Technology on Goods, Services, and Business Transactions* (pp. 148–159). Hershey, PA: IGI Global. doi:10.4018/978-1-4666-4510-3.ch008

Cheng, V., Rhodes, J., & Lok, P. (2015). The Relationship between Online Reviews, Brand Trust, and Willingness to Buy. In M. Khosrow-Pour (Ed.), *Strategic E-Commerce Systems and Tools for Competing in the Digital Marketplace* (pp. 139–161). Hershey, PA: IGI Global. doi:10.4018/978-1-4666-8133-0.ch007

Clouse, C. (2016). Design for Autonomy: Water Resources in Ladakh. In A. Goswami & A. Mishra (Eds.), *Economic Modeling, Analysis, and Policy for Sustainability* (pp. 250–265). Hershey, PA: IGI Global. doi:10.4018/978-1-5225-0094-0.ch013

Coady, D., & Hanedar, E. (2016). The Distributional Impact of Efficient Energy Pricing in India. In A. Goswami & A. Mishra (Eds.), *Economic Modeling, Analysis, and Policy for Sustainability* (pp. 194–205). Hershey, PA: IGI Global. doi:10.4018/978-1-5225-0094-0.ch010

Czyżyk, D. (2014). Chile under the Government of Sebastián Piñera. In B. Christiansen (Ed.), *Handbook of Research on Economic Growth and Technological Change in Latin America* (pp. 230–249). Hershey, PA: IGI Global. doi:10.4018/978-1-4666-6224-7.ch013

Daidj, N. (2015). *Developing Strategic Business Models and Competitive Advantage in the Digital Sector* (pp. 1–377). Hershey, PA: IGI Global. doi:10.4018/978-1-4666-6513-2

Das, R. (2014). A Game Theoretic Approach to Corporate Lending by the Banks in India. In B. Christiansen & M. Basilgan (Eds.), *Economic Behavior, Game Theory, and Technology in Emerging Markets* (pp. 271–288). Hershey, PA: IGI Global. doi:10.4018/978-1-4666-4745-9.ch015

Das, S. (2016). Consumer Power Subsidies: Brewing Crisis in India's Economy. In A. Goswami & A. Mishra (Eds.), *Economic Modeling, Analysis, and Policy for Sustainability* (pp. 145–176). Hershey, PA: IGI Global. doi:10.4018/978-1-5225-0094-0.ch008

Datta, P. K., Tiwary, H., & Chakrabarti, S. (2017). MGNREGS of India: Complementarities Between Employment and Infrastructure. In R. Das (Ed.), *Social, Health, and Environmental Infrastructures for Economic Growth* (pp. 57–81). Hershey, PA: IGI Global. doi:10.4018/978-1-5225-2364-2.ch004

Dunn, L., & Dunn, R. (2014). Natural Resources and Welfare: A Study of U.S. States. In P. Schaeffer & E. Kouassi (Eds.), *Econometric Methods for Analyzing Economic Development* (pp. 235–261). Hershey, PA: IGI Global. doi:10.4018/978-1-4666-4329-1.ch014

Erdoğdu, M. M., & Karaca, C. (2014). A Road Map for a Domestic Wind Turbine Manufacturıng Industry in Turkey. In B. Christiansen & M. Basilgan (Eds.), *Economic Behavior, Game Theory, and Technology in Emerging Markets* (pp. 57–90). Hershey, PA: IGI Global. doi:10.4018/978-1-4666-4745-9.ch005

Esmaeili, L., & Afzali, G. A. (2014). Improving Network-Based Marketing by Personalized Recommendation. In I. Lee (Ed.), *Trends in E-Business, E-Services, and E-Commerce: Impact of Technology on Goods, Services, and Business Transactions* (pp. 160–174). Hershey, PA: IGI Global. doi:10.4018/978-1-4666-4510-3.ch009

Fiodendji, K., Kamgnia, B. D., & Tanimoune, N. A. (2014). Inflation and Economic Performance in the CFA Franc Zone: Transmission Channels and Threshold Effects. In P. Schaeffer & E. Kouassi (Eds.), *Econometric Methods for Analyzing Economic Development* (pp. 10–29). Hershey, PA: IGI Global. doi:10.4018/978-1-4666-4329-1.ch002

Fornes, G., & Altamira, M. (2014). Setting the Scene for the Development of Differentiation Strategies in Emerging Markets: The Case of Chinese MNCs in Latin America. In B. Christiansen (Ed.), *Handbook of Research on Economic Growth and Technological Change in Latin America* (pp. 250–264). Hershey, PA: IGI Global. doi:10.4018/978-1-4666-6224-7.ch014

Francesconi, A., & Dossena, C. (2014). Online Corporate Reputation Management and IT: From Reactive to Proactive Organizations. In I. Lee (Ed.), *Trends in E-Business, E-Services, and E-Commerce: Impact of Technology on Goods, Services, and Business Transactions* (pp. 206–228). Hershey, PA: IGI Global. doi:10.4018/978-1-4666-4510-3.ch012

Frutos-Bencze, D. (2014). Impact of CAFTA-DR's Environmental Provisions on Member Countries and Firm-Level Environmental Voluntary Mechanisms. In B. Christiansen (Ed.), *Handbook of Research on Economic Growth and Technological Change in Latin America* (pp. 265–290). Hershey, PA: IGI Global. doi:10.4018/978-1-4666-6224-7.ch015

Garita, M. (2014). The Rationality of Dumping: The Case of Guatemala. In B. Christiansen & M. Basilgan (Eds.), *Economic Behavior, Game Theory, and Technology in Emerging Markets* (pp. 359–367). Hershey, PA: IGI Global. doi:10.4018/978-1-4666-4745-9.ch019

Geda, A., Huizinga, F., & Yimer, A. (2016). Exogenous Shocks and Macroeconomic Policy Analysis using Applied Macro-Econometric Models in Africa: The Case of Rwanda. In A. Goswami & A. Mishra (Eds.), *Economic Modeling, Analysis, and Policy for Sustainability* (pp. 74–129). Hershey, PA: IGI Global. doi:10.4018/978-1-5225-0094-0.ch006

Ghosh, D., & Dinda, S. (2017). Health Infrastructure and Economic Development in India. In R. Das (Ed.), *Social, Health, and Environmental Infrastructures for Economic Growth* (pp. 99–119). Hershey, PA: IGI Global. doi:10.4018/978-1-5225-2364-2.ch006

Ghosh, N., & Hazra, S. (2016). Sensitivity Analysis with Calibration of Natural Resource Variables under Climate Change: Comparing Computable General Equilibrium (CGE) and Econometric Frameworks. In A. Goswami & A. Mishra (Eds.), *Economic Modeling, Analysis, and Policy for Sustainability* (pp. 1–11). Hershey, PA: IGI Global. doi:10.4018/978-1-5225-0094-0.ch001

Ghosh, R. K. (2016). Modeling Transaction Costs for a Sustainable Energy Policy: A Review and Illustration. In A. Goswami & A. Mishra (Eds.), *Economic Modeling, Analysis, and Policy for Sustainability* (pp. 130–144). Hershey, PA: IGI Global. doi:10.4018/978-1-5225-0094-0.ch007

Ghosh, S. (2017). Changes in Infrastructural Condition of Slums in India: A State Level Analysis. In R. Das (Ed.), *Social, Health, and Environmental Infrastructures for Economic Growth* (pp. 14–38). Hershey, PA: IGI Global. doi:10.4018/978-1-5225-2364-2.ch002

Goswami, A., Chaudhury, S., & Garg, T. (2016). Impact of Green Growth and Development Path for Skilled and Unskilled Job Creation and Economic, Social Sustainability: Case Study of India – A Recursive Dynamic CGE Model Approach. In A. Goswami & A. Mishra (Eds.), *Economic Modeling, Analysis, and Policy for Sustainability* (pp. 29–40). Hershey, PA: IGI Global. doi:10.4018/978-1-5225-0094-0.ch003

Goswami, A., Mishra, A., & Bandopadhyay, K. (2016). Nonlinearity of Energy Transition in India and Implications for Sustainability Science: An Exploratory Insight. In A. Goswami & A. Mishra (Eds.), *Economic Modeling, Analysis, and Policy for Sustainability* (pp. 206–233). Hershey, PA: IGI Global. doi:10.4018/978-1-5225-0094-0.ch011

Hailu, Y. G., Adelaja, A., Akaeze, H., & Hanson, S. (2014). Explaining International Land Transactions in Africa. In P. Schaeffer & E. Kouassi (Eds.), *Econometric Methods for Analyzing Economic Development* (pp. 111–130). Hershey, PA: IGI Global. doi:10.4018/978-1-4666-4329-1.ch008

Hartmann, A. M., & Davila, A. (2014). Ideologies and Practices of Management in Latin America. In B. Christiansen (Ed.), *Handbook of Research on Economic Growth and Technological Change in Latin America* (pp. 19–43). Hershey, PA: IGI Global. doi:10.4018/978-1-4666-6224-7.ch002

Heinonen, J. (2016). Conjoint Analysis with fMRI: A Novel Analytical Approach to Neuromarketing. In B. Christiansen & E. Lechman (Eds.), *Neuroeconomics and the Decision-Making Process* (pp. 147–162). Hershey, PA: IGI Global. doi:10.4018/978-1-4666-9989-2.ch009

Hernández-Catá, E. (2014). Accounting for the Growth of Real GDP in Cuba, 1990-2010: An Exploratory Empirical Study. In B. Christiansen & M. Basilgan (Eds.), *Economic Behavior, Game Theory, and Technology in Emerging Markets* (pp. 18–29). Hershey, PA: IGI Global. doi:10.4018/978-1-4666-4745-9.ch002

Hernández-Catá, E. (2014). Macroeconomic Policy in Cuba: History, Problems, and the Agenda for the Future. In B. Christiansen (Ed.), *Handbook of Research on Economic Growth and Technological Change in Latin America* (pp. 326–338). Hershey, PA: IGI Global. doi:10.4018/978-1-4666-6224-7.ch018

Jackson, R. W., & Court, C. D. (2014). Embedding New Technologies and Extending Time Horizons in Input-Output Analysis. In P. Schaeffer & E. Kouassi (Eds.), *Econometric Methods for Analyzing Economic Development* (pp. 192–204). Hershey, PA: IGI Global. doi:10.4018/978-1-4666-4329-1.ch012

Jaffe, K. (2014). The Relevance of Science in Development: Scientific Development Favors Economic Prosperity, but Not Necessarily through its Effect on Technological Knowledge in Middle Income Countries. In B. Christiansen & M. Basilgan (Eds.), *Economic Behavior, Game Theory, and Technology in Emerging Markets* (pp. 1–17). Hershey, PA: IGI Global. doi:10.4018/978-1-4666-4745-9.ch001

Jain, N., Ahuja, V., & Medury, Y. (2015). Digital Marketing Optimization. In M. Khosrow-Pour (Ed.), *Strategic E-Commerce Systems and Tools for Competing in the Digital Marketplace* (pp. 162–170). Hershey, PA: IGI Global. doi:10.4018/978-1-4666-8133-0.ch008

Jana, S. K., & Karmakar, A. K. (2017). Infrastructure, Education, and Economic Development in India: A State Level Analysis. In R. Das (Ed.), *Social, Health, and Environmental Infrastructures for Economic Growth* (pp. 1–13). Hershey, PA: IGI Global. doi:10.4018/978-1-5225-2364-2.ch001

Juan Rock, T., & Pesakovic, G. (2014). Lessons Learned from Chilean Model of Innovation and Development. In B. Christiansen (Ed.), *Handbook of Research on Economic Growth and Technological Change in Latin America* (pp. 350–363). Hershey, PA: IGI Global. doi:10.4018/978-1-4666-6224-7.ch020

Karaca, C., & Erdoğdu, M. M. (2014). Sustainable Development and Turkey's Biomass Energy Potential. In B. Christiansen & M. Basilgan (Eds.), *Economic Behavior, Game Theory, and Technology in Emerging Markets* (pp. 335–358). Hershey, PA: IGI Global. doi:10.4018/978-1-4666-4745-9.ch018

Karmaoui, A. (2016). Economic Value of Regulating and Cultural Ecosystem Services of Oasean Biome: Case Study in Middle Draa Valley. In A. Goswami & A. Mishra (Eds.), *Economic Modeling, Analysis, and Policy for Sustainability* (pp. 307–326). Hershey, PA: IGI Global. doi:10.4018/978-1-5225-0094-0.ch016

Karnaukhova, O., & Nekrasova, I. (2016). Plasticity and Memory in the Financial Markets. In B. Christiansen & E. Lechman (Eds.), *Neuroeconomics and the Decision-Making Process* (pp. 163–182). Hershey, PA: IGI Global. doi:10.4018/978-1-4666-9989-2.ch010

Kasemsap, K. (2016). The Fundamentals of Neuroeconomics. In B. Christiansen & E. Lechman (Eds.), *Neuroeconomics and the Decision-Making Process* (pp. 1–32). Hershey, PA: IGI Global. doi:10.4018/978-1-4666-9989-2.ch001

Kittilaksanawong, W. (2016). Managerial Reactions to Ambiguous Environmental Changes: Attention, Reasoning, and Erratic Decisions. In B. Christiansen & E. Lechman (Eds.), *Neuroeconomics and the Decision-Making Process* (pp. 108–121). Hershey, PA: IGI Global. doi:10.4018/978-1-4666-9989-2.ch006

Knight, J. (2014). The Role and Challenges of Consumer Research in the Dynamic Economic and Social Environment of Latin America. In B. Christiansen (Ed.), *Handbook of Research on Economic Growth and Technological Change in Latin America* (pp. 174–193). Hershey, PA: IGI Global. doi:10.4018/978-1-4666-6224-7.ch010

Koumparoulis, D. N. (2015). Financial Development and Economic Growth. In N. Ray, D. Das, S. Chaudhuri, & A. Ghosh (Eds.), *Strategic Infrastructure Development for Economic Growth and Social Change* (pp. 24–37). Hershey, PA: IGI Global. doi:10.4018/978-1-4666-7470-7.ch003

Kuo, N., & Dai, Y. (2015). Predicting Low-Carbon Travel Behavior: A Modified Theory of Planned Behavior Model from Taiwan. In M. Khosrow-Pour (Ed.), *Strategic E-Commerce Systems and Tools for Competing in the Digital Marketplace* (pp. 95–116). Hershey, PA: IGI Global. doi:10.4018/978-1-4666-8133-0.ch005

Lahmiri, S. (2016). Economic Decision Making, Emotion, and Prefrontal Cortex. In B. Christiansen & E. Lechman (Eds.), *Neuroeconomics and the Decision-Making Process* (pp. 122–131). Hershey, PA: IGI Global. doi:10.4018/978-1-4666-9989-2.ch007

Lavassani, K. M., Movahedi, B., & Kumar, V. (2014). From Integration to Social Media: Understanding Electronic Marketplace. In I. Lee (Ed.), *Trends in E-Business, E-Services, and E-Commerce: Impact of Technology on Goods, Services, and Business Transactions* (pp. 1–19). Hershey, PA: IGI Global. doi:10.4018/978-1-4666-4510-3.ch001

Le Dinh, T., & Thi, T. T. (2014). Collaborative Business Service Modelling and Improving: An Information-Driven Approach. In I. Lee (Ed.), *Trends in E-Business, E-Services, and E-Commerce: Impact of Technology on Goods, Services, and Business Transactions* (pp. 128–147). Hershey, PA: IGI Global. doi:10.4018/978-1-4666-4510-3.ch007

Lechman, E. (2014). Does Technology Adoption Matter for Economic Development?: Empirical Evidence for Latin American Countries. In B. Christiansen (Ed.), *Handbook of Research on Economic Growth and Technological Change in Latin America* (pp. 1–18). Hershey, PA: IGI Global. doi:10.4018/978-1-4666-6224-7.ch001

Leng, H. K. (2014). An Update on the Use of Facebook as a Marketing Tool by Private Educational Institutions in Singapore. In I. Lee (Ed.), *Trends in E-Business, E-Services, and E-Commerce: Impact of Technology on Goods, Services, and Business Transactions* (pp. 191–205). Hershey, PA: IGI Global. doi:10.4018/978-1-4666-4510-3.ch011

Lin, C., & Jalleh, G. (2015). Evaluation of B2B Pharmaceutical Supply Chain in Australia. In M. Khosrow-Pour (Ed.), *Strategic E-Commerce Systems and Tools for Competing in the Digital Marketplace* (pp. 56–79). Hershey, PA: IGI Global. doi:10.4018/978-1-4666-8133-0.ch003

Maes, K., De Haes, S., & Van Grembergen, W. (2015). A Literature Review on IT Value Management: Assisting Organisations to Realise Benefits from IT-Enabled Investments. In M. Khosrow-Pour (Ed.), *Strategic E-Commerce Systems and Tools for Competing in the Digital Marketplace* (pp. 26–55). Hershey, PA: IGI Global. doi:10.4018/978-1-4666-8133-0.ch002

Manzoor, A. (2017). Sustainable Infrastructures: A New Infrastructure Investment Strategy. In R. Das (Ed.), *Social, Health, and Environmental Infrastructures for Economic Growth* (pp. 145–164). Hershey, PA: IGI Global. doi:10.4018/978-1-5225-2364-2.ch008

Marques, J. (2014). The Caribbean Community: A Promise yet Unfulfilled. In B. Christiansen (Ed.), *Handbook of Research on Economic Growth and Technological Change in Latin America* (pp. 125–137). Hershey, PA: IGI Global. doi:10.4018/978-1-4666-6224-7.ch007

Martins, R. A., Kumar, K., Mukherjee, A., Nabin, M. H., & Bhattacharya, S. (2014). Decision-Making in Economics: Critical Lessons from Neurobiology. In B. Christiansen & M. Basilgan (Eds.), *Economic Behavior, Game Theory, and Technology in Emerging Markets* (pp. 46–56). Hershey, PA: IGI Global. doi:10.4018/978-1-4666-4745-9.ch004

Mathews, D., & Castro, S. (2014). The Central American Clothing Assembly Industry and the Asian Competition. In B. Christiansen (Ed.), *Handbook of Research on Economic Growth and Technological Change in Latin America* (pp. 313–325). Hershey, PA: IGI Global. doi:10.4018/978-1-4666-6224-7.ch017

Matuska, E., & Landowska, A. (2016). Cooperation as a Core Component of Intercultural Competence: A Neuroeconomic Perspective. In B. Christiansen & E. Lechman (Eds.), *Neuroeconomics and the Decision-Making Process* (pp. 84–107). Hershey, PA: IGI Global. doi:10.4018/978-1-4666-9989-2.ch005

McGee, R. W. (2014). Attitudes toward Bribery: A Case Study of Brazil. In B. Christiansen (Ed.), *Handbook of Research on Economic Growth and Technological Change in Latin America* (pp. 394–416). Hershey, PA: IGI Global. doi:10.4018/978-1-4666-6224-7.ch023

McGee, R. W. (2014). The Ethics of Tax Evasion: A Case Study of Brazil. In B. Christiansen (Ed.), *Handbook of Research on Economic Growth and Technological Change in Latin America* (pp. 374–393). Hershey, PA: IGI Global. doi:10.4018/978-1-4666-6224-7.ch022

Mduma, J. K. (2014). Point Density Estimation of Changes in Income Polarization in Tanzania, 1992–2001. In P. Schaeffer & E. Kouassi (Eds.), *Econometric Methods for Analyzing Economic Development* (pp. 50–65). Hershey, PA: IGI Global. doi:10.4018/978-1-4666-4329-1.ch004

Michieka, N. M. (2014). Population Dynamics, Economic Growth, and Energy Consumption in Kenya. In P. Schaeffer & E. Kouassi (Eds.), *Econometric Methods for Analyzing Economic Development* (pp. 178–191). Hershey, PA: IGI Global. doi:10.4018/978-1-4666-4329-1.ch011

Mukherjee, S. (2017). Anatomy and Significance of Public Healthcare Expenditure and Economic Growth Nexus in India: Its Implications for Public Health Infrastructure Thereof. In R. Das (Ed.), *Social, Health, and Environmental Infrastructures for Economic Growth* (pp. 120–144). Hershey, PA: IGI Global. doi:10.4018/978-1-5225-2364-2.ch007

Mukhopadhyay, D., & Mandal, A. K. (2017). Infrastructure Development and Changing Market Penetration of Consumer Durables in Rural India: An Empirical Investigation. In R. Das (Ed.), *Social, Health, and Environmental Infrastructures for Economic Growth* (pp. 269–289). Hershey, PA: IGI Global. doi:10.4018/978-1-5225-2364-2.ch013

Mukhopadhyay, P., Sinha, M., & Sengupta, P. P. (2017). Importance of Sustainable Rural Development through Agrarian Reforms: An Indian Scenario. In R. Das (Ed.), *Social, Health, and Environmental Infrastructures for Economic Growth* (pp. 290–306). Hershey, PA: IGI Global. doi:10.4018/978-1-5225-2364-2.ch014

Muriithi, M. K., & Mwabu, G. (2014). Demand for Health Care in Kenya: The Effects of Information about Quality. In P. Schaeffer & E. Kouassi (Eds.), *Econometric Methods for Analyzing Economic Development* (pp. 102–110). Hershey, PA: IGI Global. doi:10.4018/978-1-4666-4329-1.ch007

N'Guessan, C. F. (2014). Asymmetric Equilibrium Adjustment between Employment and Economic Growth in Côte d'Ivoire: A Consistent Momentum Threshold Autoregressive Model. In P. Schaeffer & E. Kouassi (Eds.), *Econometric Methods for Analyzing Economic Development* (pp. 1–9). Hershey, PA: IGI Global. doi:10.4018/978-1-4666-4329-1.ch001

Najar, T., & Amami, M. (2014). Interorganizational Information Systems and Interorganizational Relationships. In B. Christiansen & M. Basilgan (Eds.), *Economic Behavior, Game Theory, and Technology in Emerging Markets* (pp. 186–212). Hershey, PA: IGI Global. doi:10.4018/978-1-4666-4745-9.ch011

Najar, T., & Amami, M. (2014). Interorganizational Relationships: Theoretical Evolution and Success Attributes. In B. Christiansen & M. Basilgan (Eds.), *Economic Behavior, Game Theory, and Technology in Emerging Markets* (pp. 169–185). Hershey, PA: IGI Global. doi:10.4018/978-1-4666-4745-9.ch010

Nembua, C. C. (2014). Measuring and Explaining Economic Inequality: An Extension of the Gini Coefficient. In P. Schaeffer & E. Kouassi (Eds.), *Econometric Methods for Analyzing Economic Development* (pp. 87–101). Hershey, PA: IGI Global. doi:10.4018/978-1-4666-4329-1.ch006

Nikitaeva, A. Y. (2016). Objective and Subjective Aspects of Decision-Making Support at the Mesoeconomic Level. In B. Christiansen & E. Lechman (Eds.), *Neuroeconomics and the Decision-Making Process* (pp. 183–215). Hershey, PA: IGI Global. doi:10.4018/978-1-4666-9989-2.ch011

Nondo, C., Kahsai, M. S., & Schaeffer, P. V. (2014). The Role of Governance in Teledensity and Economic Growth: GMM Estimation. In P. Schaeffer & E. Kouassi (Eds.), *Econometric Methods for Analyzing Economic Development* (pp. 262–278). Hershey, PA: IGI Global. doi:10.4018/978-1-4666-4329-1.ch015

Oyekale, A. S., Adeoti, A. I., & Oyekale, T. O. (2014). Sources of Income Inequality in Nigeria: Decomposition Approaches. In P. Schaeffer & E. Kouassi (Eds.), *Econometric Methods for Analyzing Economic Development* (pp. 66–86). Hershey, PA: IGI Global. doi:10.4018/978-1-4666-4329-1.ch005

Pal, B. D., & Pohit, S. (2016). Prioritizing Sectors for Sustainable Growth in India: An Environmental Social Accounting Matrix based Analysis. In A. Goswami & A. Mishra (Eds.), *Economic Modeling, Analysis, and Policy for Sustainability* (pp. 41–53). Hershey, PA: IGI Global. doi:10.4018/978-1-5225-0094-0.ch004

Patra, S. (2017). Energy Challenges and Infrastructure Development in South Asia. In R. Das (Ed.), *Social, Health, and Environmental Infrastructures for Economic Growth* (pp. 165–187). Hershey, PA: IGI Global. doi:10.4018/978-1-5225-2364-2.ch009

Pennington, R. (2015). Branding Cultural Analogues in Virtual Communities. In M. Khosrow-Pour (Ed.), *Strategic E-Commerce Systems and Tools for Competing in the Digital Marketplace* (pp. 117–138). Hershey, PA: IGI Global. doi:10.4018/978-1-4666-8133-0.ch006

Pessa, E. (2016). Neural Network Modeling for Organizational Psychology. In L. Biggiero, P. Angelini, M. Basevi, N. Carbonara, A. Mastrogiorgio, E. Pessa, ... M. Valente (Eds.), *Relational Methodologies and Epistemology in Economics and Management Sciences* (pp. 300–312). Hershey, PA: IGI Global. doi:10.4018/978-1-4666-9770-6.ch011

Pessa, E. (2016). Neural Network Models: Usefulness and Limitations. In L. Biggiero, P. Angelini, M. Basevi, N. Carbonara, A. Mastrogiorgio, E. Pessa, ... M. Valente (Eds.), *Relational Methodologies and Epistemology in Economics and Management Sciences* (pp. 100–127). Hershey, PA: IGI Global. doi:10.4018/978-1-4666-9770-6.ch003

Pradhan, B. K., & Ghosh, J. (2016). A CGE Analysis of the Effects of Global Climate Change Mitigation Policies on India. In A. Goswami & A. Mishra (Eds.), *Economic Modeling, Analysis, and Policy for Sustainability* (pp. 12–28). Hershey, PA: IGI Global. doi:10.4018/978-1-5225-0094-0.ch002

Ramirez, J. (2014). Entrepreneurs' Responses to Illegitimate Institutional Pressures in Monterrey, Mexico. In B. Christiansen (Ed.), *Handbook of Research on Economic Growth and Technological Change in Latin America* (pp. 194–208). Hershey, PA: IGI Global. doi:10.4018/978-1-4666-6224-7.ch011

Ramirez, R. S. (2014). Integration and Foreign Investment in Latin America. In B. Christiansen (Ed.), *Handbook of Research on Economic Growth and Technological Change in Latin America* (pp. 105–124). Hershey, PA: IGI Global. doi:10.4018/978-1-4666-6224-7.ch006

Robledo-Ardila, C., & Velasquez-Montoya, M. (2014). Inducascos S.A.: International Operations for Local Market Leadership. In B. Christiansen (Ed.), *Handbook of Research on Economic Growth and Technological Change in Latin America* (pp. 364–373). Hershey, PA: IGI Global. doi:10.4018/978-1-4666-6224-7.ch021

Rusko, R. (2016). The Role of the Mixed Strategies and Selective Inflexibility in the Repeated Games of Business: Multiple Case Study Analysis. In B. Christiansen & E. Lechman (Eds.), *Neuroeconomics and the Decision-Making Process* (pp. 132–146). Hershey, PA: IGI Global. doi:10.4018/978-1-4666-9989-2.ch008

Sayin, H. C., & Çakan, S. (2014). International Stock Investment Portfolio Management Strategies for Emerging Economies. In B. Christiansen & M. Basilgan (Eds.), *Economic Behavior, Game Theory, and Technology in Emerging Markets* (pp. 368–384). Hershey, PA: IGI Global. doi:10.4018/978-1-4666-4745-9.ch020

Seçilmiş, İ. E. (2014). A Survey of Game Theory Applications in Turkey. In B. Christiansen & M. Basilgan (Eds.), *Economic Behavior, Game Theory, and Technology in Emerging Markets* (pp. 155–168). Hershey, PA: IGI Global. doi:10.4018/978-1-4666-4745-9.ch009

Şen, Z. (2014). Intelligent Business Decision-Making in Global Organizations via Fuzzy Reasoning Systems. In B. Christiansen & M. Basilgan (Eds.), *Economic Behavior, Game Theory, and Technology in Emerging Markets* (pp. 127–154). Hershey, PA: IGI Global. doi:10.4018/978-1-4666-4745-9.ch008

Sharma, S. (2016). Economic Damage Due to Ozone Pollution in NCR: Ozone Impacts. In A. Goswami & A. Mishra (Eds.), *Economic Modeling, Analysis, and Policy for Sustainability* (pp. 284–306). Hershey, PA: IGI Global. doi:10.4018/978-1-5225-0094-0.ch015

Smith, A. D. (2014). Beneficial E-Personalization and Related Technological Innovations Applied to E-Tailing. In I. Lee (Ed.), *Trends in E-Business, E-Services, and E-Commerce: Impact of Technology on Goods, Services, and Business Transactions* (pp. 20–46). Hershey, PA: IGI Global. doi:10.4018/978-1-4666-4510-3.ch002

Tian, Z., Zhang, Z., & Guan, X. (2015). A General Evolution Mechanism Model for E-Commerce Network. In M. Khosrow-Pour (Ed.), *Strategic E-Commerce Systems and Tools for Competing in the Digital Marketplace* (pp. 80–94). Hershey, PA: IGI Global. doi:10.4018/978-1-4666-8133-0.ch004

Tran, B. (2014). Game Theory vs. Business Ethics: The Game of Ethics. In B. Christiansen & M. Basilgan (Eds.), *Economic Behavior, Game Theory, and Technology in Emerging Markets* (pp. 213–236). Hershey, PA: IGI Global. doi:10.4018/978-1-4666-4745-9.ch012

Tran, B. (2014). The Evolution of Business Ethics to Business Law. In B. Christiansen & M. Basilgan (Eds.), *Economic Behavior, Game Theory, and Technology in Emerging Markets* (pp. 109–126). Hershey, PA: IGI Global. doi:10.4018/978-1-4666-4745-9.ch007

Tuncel, C. O. (2014). Sectoral System of Innovation and Technological Upgrading Strategies: A Comparative Study of the Automotive Industry in South Korea and Turkey. In B. Christiansen & M. Basilgan (Eds.), *Economic Behavior, Game Theory, and Technology in Emerging Markets* (pp. 289–322). Hershey, PA: IGI Global. doi:10.4018/978-1-4666-4745-9.ch016

Tuncel, C. O., & Gürsel, V. (2014). Significance of Innovation Policies: Lessons for Countries Caught in the Middle Income Trap. In B. Christiansen & M. Basilgan (Eds.), *Economic Behavior, Game Theory, and Technology in Emerging Markets* (pp. 323–334). Hershey, PA: IGI Global. doi:10.4018/978-1-4666-4745-9.ch017

Valente, M. (2016). Agent-Based Simulation Modeling: Definitions and a Methodological Proposal. In L. Biggiero, P. Angelini, M. Basevi, N. Carbonara, A. Mastrogiorgio, E. Pessa, ... M. Valente (Eds.), *Relational Methodologies and Epistemology in Economics and Management Sciences* (pp. 128–139). Hershey, PA: IGI Global. doi:10.4018/978-1-4666-9770-6.ch004

Valente, M. (2016). Innovations and Organizational Structures. In L. Biggiero, P. Angelini, M. Basevi, N. Carbonara, A. Mastrogiorgio, E. Pessa, ... M. Valente (Eds.), *Relational Methodologies and Epistemology in Economics and Management Sciences* (pp. 283–299). Hershey, PA: IGI Global. doi:10.4018/978-1-4666-9770-6.ch010

Vaquero, M. G., & Saiz-Alvarez, J. M. (2016). Smart Cities in Spain – Policy, Sustainability, and the National Plan: New Political Measures, Agents, and Sustainability. In A. Goswami & A. Mishra (Eds.), *Economic Modeling, Analysis, and Policy for Sustainability* (pp. 266–283). Hershey, PA: IGI Global. doi:10.4018/978-1-5225-0094-0.ch014

Vargas-Hernández, J. G., Muratalla-Bautista, G., & Cruz, I. D. (2017). Point of View of Economical Organization at Manzanillo's Harbor. In R. Das (Ed.), *Social, Health, and Environmental Infrastructures for Economic Growth* (pp. 307–324). Hershey, PA: IGI Global. doi:10.4018/978-1-5225-2364-2.ch015

Varma, A., & Singhal, S. (2016). Significance of Cultural Dimensions in Urban Planning Process for Sustainability of Pilgrimage Towns in India. In A. Goswami & A. Mishra (Eds.), *Economic Modeling, Analysis, and Policy for Sustainability* (pp. 327–350). Hershey, PA: IGI Global. doi:10.4018/978-1-5225-0094-0.ch017

Virkar, S. (2016). Successfully Playing Games of Tax Compliance: Government Agencies as Architects of Public Choice. In B. Christiansen & E. Lechman (Eds.), *Neuroeconomics and the Decision-Making Process* (pp. 216–245). Hershey, PA: IGI Global. doi:10.4018/978-1-4666-9989-2.ch012

Wan-Ismail, W. A., Zakaria, N., & Abdul-Talib, A. (2016). Craving vs. Compulsion for Luxury Goods?: Trends and Patterns of Conspicuous Consumption Behavior in Asian Culture. In B. Christiansen & E. Lechman (Eds.), *Neuroeconomics and the Decision-Making Process* (pp. 287–297). Hershey, PA: IGI Global. doi:10.4018/978-1-4666-9989-2.ch015

West, G. R., & Jackson, R. W. (2014). Simulating Impacts on Regional Economies: A Modeling Alternative. In P. Schaeffer & E. Kouassi (Eds.), *Econometric Methods for Analyzing Economic Development* (pp. 132–152). Hershey, PA: IGI Global. doi:10.4018/978-1-4666-4329-1.ch009

Windsor, D. (2014). Business Ethics in Emerging Economies: Identifying Game-Theoretic Insights for Key Issues. In B. Christiansen & M. Basilgan (Eds.), *Economic Behavior, Game Theory, and Technology in Emerging Markets* (pp. 30–45). Hershey, PA: IGI Global. doi:10.4018/978-1-4666-4745-9.ch003

Yaşgül, Y. S. (2014). A Structurationist Analysis of Differing Policies Oriented to the Pharmaceutical Industry: Turkey and Brazil. In B. Christiansen (Ed.), *Handbook of Research on Economic Growth and Technological Change in Latin America* (pp. 291–312). Hershey, PA: IGI Global. doi:10.4018/978-1-4666-6224-7.ch016

Youssef, M. A. (2015). Electronic Commerce and Change in Management Accounting Practices in an Egyptian Organization. In M. Khosrow-Pour (Ed.), *Strategic E-Commerce Systems and Tools for Competing in the Digital Marketplace* (pp. 189–205). Hershey, PA: IGI Global. doi:10.4018/978-1-4666-8133-0.ch010

Zapata, L. F., & Hinestroza, I. C. (2014). Technological Change and Innovation in Latin American Emerging Economies: The Pork Industry of Antioquia, Colombia. In B. Christiansen (Ed.), *Handbook of Research on Economic Growth and Technological Change in Latin America* (pp. 138–153). Hershey, PA: IGI Global. doi:10.4018/978-1-4666-6224-7.ch008

Zhang, G., Li, C., Zhang, Y., Xing, C., Xue, S., & Liu, Y. (2015). Semantic++ Electronic Commerce Architecture and Models in Cloud. In M. Khosrow-Pour (Ed.), *Strategic E-Commerce Systems and Tools for Competing in the Digital Marketplace* (pp. 1–25). Hershey, PA: IGI Global. doi:10.4018/978-1-4666-8133-0.ch001

Zhang, J., Zhong, W., & Mei, S. (2015). Purchase-Based Targeted Advertising: A Competitive Analysis. In M. Khosrow-Pour (Ed.), *Strategic E-Commerce Systems and Tools for Competing in the Digital Marketplace* (pp. 171–188). Hershey, PA: IGI Global. doi:10.4018/978-1-4666-8133-0.ch009

Index

B

BD Decision Intelligence Pipeline 78-79
Big Data 72
Business Model and Leadership Platform 50
Business Platform 2-3, 5, 11, 25, 45-46, 49, 52-53, 58, 63-64, 72-73
Business Platform Ecosystem 5

C

Customer Experience Platform 56-57

D

Data and Analytics Platform 58-59, 72
Delivery platform 52
Digital Business Platform 45-46, 49, 53, 58, 63-64, 72-73
Digital or Electronic Business (E-Business) 132

E

E-Acquiring 117, 132
Ecosystem Platform 60
E-Insurance 116, 132
Electronic Banking (E-Banking, Internet Banking, Virtual Banking or Online Banking) 133
Electronic Finance (E-Finance) 133
E-Money 116, 123, 133
E-Trading 116, 120, 133

G

Google 2, 4, 31-32, 34, 98, 101, 134-138, 140, 142-144

I

Information Systems Platform 54, 57, 59-60
Innovation Platforms 72, 96, 98, 113
Integration Platform 58, 62-63
Internet Intermediaries 24, 114, 133
Internet of Things 59, 73, 96-97
IOT MSP Business Models 106
IoT Platform 59-60, 99, 108

M

Multi-Sided Financial Platform (Henceforth, MFP) 118, 133
Multi-Sided Platform (Henceforth, MSP) 7, 133
Multi-Sided Platforms 1-2, 7, 24-25, 29-30, 36, 96, 103, 113, 133, 135, 138, 146, 159

P

Platform Building Blocks 1, 7-8, 36, 104, 150, 159-160
Platform Business Model 1, 12, 19, 24, 30, 33, 36, 51, 64-66
Platform innovation 47, 64-66, 113
Platform Intermediaries 1, 32, 34, 36, 113-114, 159

Platform Strategy 1, 9-12, 17, 19, 36, 47, 49-50, 63-64
Platform Value 8-10
Platform Value Chain 8
Promotion Platform 53

S

Search Engine 4, 31, 125, 134, 136-137, 140-141
Search Multi-sided Platforms 138
SMSP 134-135, 138, 140-143, 159
Strategy 1, 4, 9-12, 17-19, 36, 45, 47, 49-50, 53, 58, 63, 65, 73, 76, 78-79, 85, 87-88, 96, 102, 114, 119, 150

T

Talent Platform 51-52
Taxonomy 1, 13, 21-22, 30, 34, 36, 64, 85-87, 96, 98-102, 106, 108, 113-114, 117, 124, 127-128, 130, 133-134, 144, 150, 152, 159, 161
Technology Platforms 2-3, 10, 32, 53, 63-65
Transaction Platforms 113, 117
Transactional Platforms 119, 134-136, 146
Trust Platform 62

Y

Yandex 4, 121, 125, 127-128, 134, 138, 140-144

www.ingramcontent.com/pod-product-compliance
Lightning Source LLC
LaVergne TN
LVHW061533290926
849162LV00004B/784